INTERNATIONAL DEVELOPMENT IN FOCUS

Climate Afflictions

IFFAT MAHMUD, WAMEQ A. RAZA, AND MD RAFI HOSSAIN

Contents

Figures

Acknowledgments

The authors of the report are indebted to the Bangladesh Meteorological Department for sharing weather data, and particularly for the cooperation extended by Mr. Bazlur Rashid, meteorologist. The authors would like to recognize the team members of Data International who collected data for the two rounds of the panel survey, specifically Mr. Nazmul Hossain and Mr. A. F. M. Azizur Rahman. The authors gratefully acknowledge contributions of Mr. Faizuddin Ahmed, Ms. Aneire Khan, and Mr. Jyotirmoy Saha. The authors are also grateful for the collaboration extended by Dr. Syed Shabab Wahid with guidance from Professor Brandon A. Kohrt, George Washington University, for the analyses on mental health issues.

Ms. Gail Richardson, practice manager of Health, Nutrition, and Population, South Asia Region of the World Bank, provided oversight for this report, and the authors deeply appreciate her continued support and encouragement. The draft report was shared with the Climate Change and Health Promotion Unit and the Institute of Epidemiology and Disease Control Research of the Ministry of Health and Family Welfare of the Government of Bangladesh. The authors are thankful for their technical advice and collaboration.

The authors express their gratitude to the peer reviewers Mr. Stephen Geoffrey Dorey (health specialist), Ms. Anna Koziel (senior health specialist), and Mr. Muthukumara Mani (lead economist), as well as Mr. Dhushyanth Raju (lead economist), Dr. Shiyong Wang (senior health specialist), and Dr. Tamer Samah Rabie (lead health specialist) for their valuable comments. The authors are grateful to Ms. Mercy Tembon, country director for Bangladesh and Bhutan, World Bank, who chaired an internal review meeting to seek expert inputs for finalization of the report.

The authors are grateful for the financial support mobilized by the Global Facility for Disaster Reduction and Recovery Multi-Donor Trust Fund as well as the Health Sector Support Project Multi-Donor Trust Fund, which is cofinanced by the Embassy of the Kingdom of the Netherlands; the Foreign, Commonwealth, and Development Office of the United Kingdom; Gavi, the Vaccine Alliance; Global Affairs Canada; and the Swedish Development Cooperation Agency.

About the Authors

Md Rafi Hossain is an operations officer with the World Bank's South Asia Health, Nutrition, and Population (HNP) Global Practice. His primary responsibilities include providing strategic, technical, and operational support to the World Bank's Human Development and HNP portfolio. In his previous role as a fiduciary specialist, Rafi was involved in the operations and fiduciary management of the HNP portfolio in Bangladesh as well as several core analytical products on health financing and health economics. Before joining the World Bank, Rafi was an economist at the Policy Research Institute of Bangladesh, where he was instrumental in some of the technical team's flagship projects, such as the drafting of the Sixth Five-Year Plan and the National Social Security Strategy. Rafi has also taught at various institutions in Bangladesh and the United States. He holds a master of science degree in finance and economics from University of London and a doctorate in economics from Northern Illinois University.

Iffat Mahmud, a public policy practitioner focusing on human development, is a senior operations officer at the World Bank's Health, Nutrition, and Population Global Practice. She has more than 15 years of experience in South Asia and Eastern Africa—particularly in Bangladesh, India, Nepal, Pakistan, Tanzania, and Uganda—leading project management and policy dialogue with governments and partners. Her expertise is in advising public sector institutions on policies, strategies, and interventions; managing projects; and developing analytics. Her work spans a wide range of themes, including maternal and child health, nutrition, communicable diseases, health service delivery, institutional reforms, systems development, and emergency response. Her work includes cross-cutting areas such as climate change, water and sanitation, and social protection. She holds a bachelor of science degree in economics and a master of science degree in management from the London School of Economics and Political Science.

Wameq A. Raza is a health and nutrition specialist with the World Bank's Health, Nutrition, and Population Global Practice in Bangladesh. Before this role, he served as a poverty economist. He is an applied microeconomist with national and international experience in analytical work, program design, and implementation across eight countries in Asia and Sub-Saharan Africa. His thematic experiences include health and nutrition, social protection, and ultrapoverty interventions. He has published peer-reviewed articles in leading journals in the economics and health fields. Wameq holds a master's degree in development economics from the University of Sussex and a doctorate in health economics from Erasmus University Rotterdam.

Executive Summary

WHY THIS REPORT?

At the outset, the report makes a clear distinction between the concepts of climate change and variability. Climate variability refers to short-term changes in the average meteorological conditions over a month, a season, or a year. Climate change, however, refers to changes in average meteorological conditions and seasonal patterns over a much longer time period (Mani and Wang 2014). With these concepts clarified, the report makes the case that the consequences related to climate change and/or climate variability are well hypothesized and documented. The expanse of literature linking climate change or climate variability and health, however, is less well established. Compared to the availability of global evidence on this topic, evidence from Bangladesh is even more limited. Among those studies available for Bangladesh, some require further substantiation, as these are mostly regional one-off studies that do not account for representativeness of the population, that use disease-specific data from hospital admission records, missing out on nonhospitalized cases, and that use climatic conditions from a time that does not always match the time period of the illness being explored.

In an effort to fill this knowledge gap, the report analyzes and presents evidence in four broad areas:

1. It systematically summarizes relevant literature on the links between climate change or climate variability and health, as well as the influence of climate variability on a mosquito's life cycle, as the insect is one of the largest sources of vector-borne diseases.

2. It quantifies the relationship between climate variability and infectious disease issues in Bangladesh using primary household-level data, representative of urban and rural areas. Urban areas are further distinguished between two large city corporations and other urban spaces.[1]

3. It measures the prevalence and depth of mental health issues (stress and depression) in the sample using globally recognized standards[2] and establishes its relationship to climate variability and seasonal patterns in Bangladesh.

4. It documents the change in weather patterns in Bangladesh over the past 44 years.

The report responds to several key questions, as summarized here in the executive summary. What it does not do is construct mathematical models for projecting incidence and prevalence of infectious diseases and mental health issues based on predicted climate change patterns; neither does it attempt to establish a causal relationship between climate change and the selected health conditions.

The report uses primary data from a nationally representative sample of around 3,600 households that were surveyed during the monsoon and dry seasons. It links weather variables, incidence of selected diseases, and health conditions in Bangladesh to ensure the findings are based on precise climate and health data as much as possible. The recommendations, therefore, are context specific and drawn from primary evidence.

Bangladesh's extreme vulnerability to the effects of climate change is well documented. Through a complex pathway, climatic conditions have already negatively impacted human health. This is likely to escalate, given the predicted changes in weather patterns. Infectious disease transmission will change in variation and incidence for certain vector-borne diseases—such as malaria and dengue—and waterborne diseases such as diarrhea and cholera. Incidences of respiratory disease will be affected by extreme temperatures that exacerbate allergens and air pollution (World Bank 2012). If global warming progresses toward a 4°C increase scenario—the worst-case scenario presented at the 2015 Paris Climate Change Conference of Practitioners—stresses on human health can overburden the country's systems to a point where adaptation will no longer be possible (World Bank 2012). Hence, it is urgent that the public sector be better prepared to respond to the crisis.

HOW DOES WEATHER AFFECT INFECTIOUS DISEASES?

Climatic conditions impact the epidemiology of infectious diseases. Furthermore, these climatic factors interact with additional factors such as behavioral, demographic, and socioeconomic ones that influence the incidence, emergence, and distribution of such infectious diseases (Watts et al. 2018). Climate suitability for climate-sensitive infectious diseases has increased globally (Watts et al. 2020). Vectorial capacity is increasing for a number of climate-sensitive diseases,[3] with exposures along a range of temperature and rainfall. These are most acutely experienced in low- and middle-income countries (Watts et al. 2019). The number of cases of dengue fever, which is spread by mosquitoes, recorded annually has doubled every decade since 1990, and one of the potential factors that contributed to this increase is climate change (Watts et al. 2020). For malaria, another mosquito-borne disease, climate suitability has remained the same for the Southeast Asia region, which includes Bangladesh.

HOW HAS THE WEATHER CHANGED IN BANGLADESH?

Over the past 44 years, Bangladesh has become hotter, with a 0.5°C increase in mean temperature recorded between 1976 and 2019, using three-year moving averages. Trend analyses indicate that the maximum temperature generally continues to rise for all months, and the increment has been the largest from February to November. Urban centers such as the cities of Dhaka and Chattogram present unique contexts in Bangladesh. Overall, summers are becoming hotter and longer, with the monsoon season being extended from February to October, while winters are also becoming warmer. With these changes, Bangladesh appears to be losing its distinct seasonal variations.

WHAT DO THE CLIMATE VARIABILITY AND SEASONAL PATTERNS MEAN FOR PREVALENCE OF DENGUE IN DHAKA?

Existing literature indicates that cases of dengue increase between the maximum temperature range of 25°C to 35°C, with a peak at 32°C. Fouque and Reeder (2019) concluded that suitable temperature conditions are factors that determine disease transmission—even if the competent mosquito species or vectors are present, the disease will not spread unless the temperature is suitable. In Dhaka city, the average monthly maximum temperature has risen above 32°C for the months of March to October, with temperatures between April and June reaching close to or above 35°C. Average rainfall is within the range of 200 to 800 millimeters that is conducive to mosquito breeding, with increasing trends noted for the months of April to August. In addition, average humidity in Dhaka city is within the range of 60 to 80 percent, which is conducive to mosquito breeding. Lower levels of humidity at higher temperatures are more conducive to mosquitoes spreading dengue to breed and reproduce. Furthermore, other compounding factors such as urbanization, travel, and demographic change that affect the spread of main vectors of dengue (Ebi and Nealon 2016) are also relevant for Dhaka city. For example, 51 percent of the total dengue cases reported in 2019 for Bangladesh were in Dhaka city while dengue-related deaths in Dhaka city accounted for 77 percent of the total deaths recorded in the country. The 2019 dengue outbreak in Dhaka could be partially explained by weather patterns—heavy rain in February, which experienced the highest ever recorded in the previous 45 years, followed by favorable temperature and humidity in the subsequent months. With falling humidity levels, rising temperatures, and increasing rainfall in the summer months, the risk of spread of dengue may be higher in Dhaka city in the future.

DOES A CHANGE IN SEASON MAKE PEOPLE SICK?

On average, the likelihood of contracting an infectious disease is 19.7 percentage points lower in the dry season than during the monsoon. If disaggregation by disease type—vector-borne, waterborne, and respiratory diseases—is considered, the trend holds for vector-borne diseases like dengue, malaria, and associated symptoms, with 25 percent of the respondents suffering from them in the monsoon season compared to 14 percent in the dry season. For waterborne diseases and respiratory illnesses, the opposite is true, with the incidence being higher in the dry season compared to monsoon.

HOW DO TEMPERATURE AND HUMIDITY LEVELS AFFECT THE SPREAD OF DISEASES?

Humidity and mean temperature are negatively correlated to waterborne diseases but positively correlated to respiratory illnesses. A 1 percent increase in relative humidity reduces the likelihood of contracting a waterborne disease by 1.6 percentage points, while an increase in 1°C mean temperature reduces its likelihood by 4.2 percentage points. For respiratory illnesses, there is a positive association with higher humidity and temperatures—for every 1 percent increase in humidity, the likelihood of a respiratory illness increases by 1.5 percentage points, and for a 1°C increase in mean temperature it increases by 5.7 percentage points. For vector-borne diseases, an increase in temperature reduces the likelihood of disease by 1.4 percentage points, although this is not statistically significant.

DO MEGACITIES EXPERIENCE A LARGER SPREAD OF INFECTION?

Irrespective of seasonality, monsoon or dry, a higher proportion of respondents residing in Dhaka and Chattogram cities reported experiencing an infectious disease compared to the averages for national, rural, and all urban areas, which include Dhaka and Chattogram cities. When disease disaggregation is considered, the proportion of incidence was higher in Dhaka and Chattogram cities (34 percent) compared to the national average (25 percent), rural areas (22 percent), and all urban areas (25 percent) in the monsoon when vector-borne diseases are more prevalent. During the dry season, when waterborne diseases and respiratory illnesses are more prevalent compared to the monsoon, the cities of Dhaka and Chattogram report more respiratory illnesses compared to other areas, possibly due to higher exposure of its residents to air pollution. The incidence for waterborne diseases in Dhaka and Chattogram cities is lower than in other areas in the dry season.

IS AGE JUST A NUMBER OR IS MORBIDITY LINKED TO IT?

Incidence of infectious diseases increases with age, across the seasons, monsoon and dry. Disaggregation by disease category reveals a different pattern. Prevalence of respiratory illnesses is the highest among the elderly—ages 65 years and above—and increases in the dry season—72 percent in monsoon and 83 percent in the dry season. Waterborne diseases are more common among children under 5 years of age. A different pattern presents for vector-borne diseases; of the various age groups, the prevalence is the highest among adults—ages 20 to 64 years—across the two seasons.

HOW IS MENTAL HEALTH FARING?

The findings in this study demonstrate nationally representative results of depression and anxiety and their determinants. Overall, 16 percent of the respondents report suffering from depression, while 6 percent report anxiety disorders. The most vulnerable for depression and anxiety are older, poorer, and disabled individuals. On another strand, while females are at higher risk than men for depression, men are more susceptible to anxiety. Residents of urban centers are generally more anxious than their rural counterparts. Further analysis of the relationship between weather and depression and anxiety suggests that while temperature is negatively correlated to depression, anxiety is elevated by increases in temperature and humidity.

WHAT DOES IT ALL MEAN?

Climate change, as currently understood, is precipitating an increase in global temperature along with extreme weather events. As corroborated by global literature and primary analysis undertaken for this report, climate variability and seasonal changes influence the prevalence of infectious diseases and affect the mental health of people. The most vulnerable are children and the elderly, as well as those living in large metropolises. With climate further predicted to change, the deleterious effects on human physical and mental health are likely to escalate. The discussions point to the need for the following:

• Improving data collection systems for improved predictability and localization of weather data, which will help in tracking the impact of climate variability on diseases
• Strengthening health systems to preempt and mitigate potential outbreaks of infectious and other emerging or reemerging climate-sensitive diseases
• Ensuring adequacy of response mechanisms for better adaptation to the effects of climate change

Both mitigation and adaptive measures need to be prioritized; otherwise the progress the country has made thus far could be eroded.

WHAT ARE THE MAIN RECOMMENDATIONS?

Based on the findings, recommendations included in this report mostly focus on measures to increase capacity to record accurate weather data at a more localized and granular level and linking it to health data. These activities will assist in (1) tracing the evolution of climate-sensitive diseases; (2) strengthening disease surveillance and establishing a climate-based dengue early warning system, which will use weather data to predict potential disease outbreaks;

(3) enhancing vector-control measures through innovative approaches; (4) addressing mental health issues through improved assessments and facilitating means to address the shortcomings; and (5) measuring air quality to tackle air pollution, which is an important compounding factor for the spread of diseases.

HOW DO THE FINDINGS AND RECOMMENDATIONS SUPPORT THE POLICY DISCOURSE?

Notwithstanding the limitations of the study, it is important to note that this report intends to establish a causal link between climate variability as well as seasonal variation and human health. The findings will assist practitioners and subject matter experts in policy dialogue that will contribute to taking forward the World Bank's corporate commitment on climate change. The policy dialogue facilitated through this document will focus on supporting governments in further developing and implementing mitigation, adaptation, and resilience measures. Finally, it is hoped that the report will pave the way for future research that focuses on building a stronger evidence base on the relationship between climate change and health and fuel the need for strengthening health systems based on evidence.

NOTES

1. Sampling also allows for disaggregated analysis for major urban centers, that is, Dhaka and Chattogram cities.
2. Depression is measured using the Patient Health Questionnaire-9 (PHQ-9), a commonly used depression screening instrument comprising nine items on a four-point Likert response scale. Anxiety is measured with the Generalized Anxiety Disorder-7 (GAD-7), a seven-item, four-point, Likert-style anxiety screening scale. Locally contextualized versions of the tools were used.
3. Vectorial capacity is defined as "a measurement of the efficiency of vector-borne disease transmission" (Norris 2006).

REFERENCES

Ebi, K. L., and J. Nealon. 2016. "Dengue in a Changing Climate." *Environmental Research* 151 (2016): 115–23.

Fouque, F., and J. C. Reeder. 2019. "Impact of Past and On-going Changes on Climate and Weather on Vector-Borne Diseases Transmission: A Look at the Evidence." *Infectious Diseases of Poverty* 8 (51).

Mani, M., and L. Wang. 2014. *Climate Change and Health Impacts: How Vulnerable Is Bangladesh and What Needs to Be Done*. Washington, DC: World Bank. https://openknowledge.worldbank.org/handle/10986/21820.

Norris, Douglas E. 2006. "Malaria Entomology." Presentation. Johns Hopkins Bloomberg School of Public Health. https://www.glowm.com/pdf/M-lecture4.pdf.

Watts, N., M. Amann, N. Arnell, S. Ayeb-Karlsson, L. Beagley, K. Belesova, M. Boykoff, et al. 2020. *Responding to Convergence Crises*. The 2020 report of the *Lancet* Countdown on health and climate change. *Lancet* 397 (10269): 129–70. doi:10.1016/S0140-6736(20)32290-X.

Watts, N., M. Amann, N. Arnell, S. Ayeb-Karlsson, K. Belesova, H. Berry, T. Bouley, et al. 2018. *Shaping the Health of Nations for Centuries to Come*. The 2018 report of the *Lancet* Countdown on health and climate change, *Lancet* 392 (10163): 2479–514. doi:10.1016/S0140-6736(18)32594-7.

Watts, N., M. Amann, N. Arnell, S. Ayeb-Karlsson, K. Belesova, M. Boykoff, P. Byass, et al. 2019. *Ensuring That the Health of a Child Born Today Is Not Defined by a Changing Climate*. The 2019 report of the *Lancet* Countdown on health and climate change, *Lancet* 394 (10211): 1836–78. doi:10.1016/S0140-6736(19)32596-6.

World Bank. 2012. *Turn Down the Heat: Why a 4°C Warmer World Must Be Avoided*. Washington, DC: World Bank. https://openknowledge.worldbank.org/handle/10986/11860.

Abbreviations

BMD	Bangladesh Meteorological Department
BMRC	Bangladesh Medical Research Council
CCHPU	Climate Change and Health Promotion Unit
DGHS	Directorate General of Health Services (Bangladesh)
EA	enumeration area
ENSO	El Niño–Southern Oscillation
GAD-7	Generalized Anxiety Disorder-7
IEDCR	Institute of Epidemiology and Disease Control Research
IPCC	Intergovernmental Panel on Climate Change
MoHFW	Ministry of Health and Family Welfare
NCD	noncommunicable disease
PHQ-9	Patient Health Questionnaire-9
PSU	primary sampling unit
PTSD	post-traumatic stress disorder
RCP	representative concentration pathway
WASH	water, sanitation, and hygiene
WHO	World Health Organization

Introduction and Overview

1 Introduction

BACKGROUND

Global climate change and increasing variability pose serious health risks to the global population. Abundant literature indicates that vulnerable populations, especially in low- and middle-income countries, are bearing the brunt disproportionately, exacerbating existing disparities in social determinants of health. The Global Climate Risk Index ranks Bangladesh as the world's seventh most affected country over the period 1999–2018.[1] Since 2000, the frequency of weather-related disasters has increased by 46 percent, and the total value of economic losses resulting from climate-related events increased substantially since 1990, totaling US$129 billion in 2016 (Watts et al., "Shaping the Health of Nations," 2018). Sixty percent of worldwide deaths caused by cyclones in the past 20 years occurred in Bangladesh (Bangladesh, MoEF 2008). The World Health Organization (WHO) estimates that 12.6 million people die each year because of climate change and pollution (WHO 2016). Some of the direct impacts of climate change include outbreaks and spread of infectious diseases, heat-stress-related mortality from extreme high temperatures, and mortality and morbidity from extreme weather events such as floods and storms; indirect effects are through its effect on the replication and spread of microbes and vectors (Mani and Wang 2014). Climate variability has already increased the survival and reproduction of mosquitoes and, consequently, the incidence of diseases spread by them. The World Bank's *Turn Down the Heat* 2012 report describes the influence of extreme weather events and changes in temperature, precipitation, and humidity on health. Infectious disease transmission will change in range and incidence for certain vector-borne diseases—malaria and dengue—and waterborne diseases—such as diarrhea and cholera—while the incidence of respiratory disease will be affected by allergens and air pollution exacerbated by extreme temperatures (World Bank 2012).

According to WHO (2003), "Globally, temperature increases of 2–3°C would increase the number of people who, in climatic terms, are at risk of malaria by around 3–5%" (17). The WHO model predicts 10 percent more diarrheal disease by 2030 than if climate change did not occur. The incidence of vector-borne diseases is likely to increase, based on an estimation that the capacity of mosquitoes to transmit dengue fever increased by 9.5 percent globally since 1950, due to changing climatic conditions in dengue-endemic countries (Watts et al., "From 25 Years of Inaction," 2018).

WHO and the United Nations Framework Convention on Climate Change (2015) project mean annual temperature to rise between 1.4°C and 4.8°C over the period 1990–2100 in Bangladesh. By 2070, it is estimated that between 117 million and 147 million people will be at risk of malaria, depending on emission levels. Their report states, "The health sector currently does not have adequate funding, infrastructure, human resource capacity, logistics and services required to fully address the impact of climate change on human health" (WHO and UNFCCC 2015, 1). The Government of Bangladesh prepared a Climate Change Strategy and Action Plan in 2008 (Bangladesh, MoEF 2008), highlighting the need to implement surveillance systems for existing and new disease risks and to ensure that health systems are prepared to meet future demands. According to the strategy document, "Climate change is likely to increase the incidence of water-borne and air-borne diseases. Bacteria, parasites, and disease vectors breed faster in warmer and wetter conditions and where there is poor drainage and sanitation"(WHO and UNFCCC 2015, 5).

According to the International Panel on Climate Change Fifth Assessment Report (IPCC 2014), it is likely that climate change has already negatively impacted health, even though the present worldwide burden of ill health from climate change is not well quantified. The link between health and climate change is weak (Lancet 2019) and hence the need for further research in the area of climate change and health (Watts et al., "Shaping the Health of Nations," 2018). In 2017, 43,000 articles were published in the general area of climate change, of which only 4 percent made any link to health, and less than 1 percent had a specific focus on health and climate change. Most of the scientific interest in health and climate change in 2017 was focused on America and Europe. Less than 10 percent of the papers related to health and climate change in 2017 were about Africa and Southeast Asia, which includes Bangladesh (Watts et al., "Shaping the Health of Nations," 2018). With climatic conditions projected to worsen, severely climate-change-affected countries like Bangladesh are likely to bear a greater brunt of the adverse effects. Hence, there is the need to understand better how the climate has changed over the years and to document its impact on human health.

Existing literature on the link between climate change and health in Bangladesh will benefit from further substantiation, as existing studies mostly use small nonrepresentative data, present findings with a narrow regional focus, or are observational studies with limited analyses. Mani and Wang (2014) reported that the health impact of climate variability differed greatly between premonsoon and monsoon seasons, based on a study on Bangladesh. Using monthly surveillance data in regions with a high incidence of vector-borne diseases, the report identified strong seasonal patterns between climate variability and vector-borne diseases but showed no clear trends over the previous decade. The authors used secondary health survey data retrofitted with historical weather data to analyze the relationship between climate variability and incidences of morbidity, which may have resulted in imprecisions in estimating this relationship. The paper reports a marginal but positive relationship between the two. Notwithstanding the findings, the authors conclude—based on a review of existing literature on health and climate change, particularly the links between climate variability and infectious diseases—that this important area of research is still in nascent stages and merits further investigation.

The combination of climate change and poverty is projected to affect between 35 million and 122 million people by 2030 (Balasubramanian 2018). If average global temperature increases by 4°C—which is the worst-case scenario for global warming presented at the Paris Climate Change Conference of Parties in 2015—stresses on human health can overburden the systems to a point where adaptation will no longer be possible (World Bank 2012). Hence, it is urgent that the public sector be better prepared to respond to the crisis. Although populations around the world are developing adaptive mechanisms like public health strategies and improved surveillance to cope with the impacts, the prevailing levels of adaptation are expected to be insufficient in the future (Watts et al., "Shaping the Health of Nations," 2018).

EFFECTS OF CLIMATE CHANGE ON HEALTH: WHY FOCUS ON SPECIFIC DISEASES?

Effects of climate change on human health can be direct and indirect and immediate or delayed (McMichael, Montgomery, and Costello 2012). The main pathways and categories of the health impacts of climate change are shown in figure 1.1. The direct or immediate effects include risks associated with increased frequency and intensity of heat waves and extreme weather events such as floods, cyclones, storm surges, droughts, and altered air quality (McMichael, Montgomery, and Costello 2012). The indirect effects occur through changes and disruptions to ecological and biophysical systems, which may result in altered food production, leading to undernutrition, water insecurity, air pollution, infectious diseases, mental health issues, and forced migration with accompanying societal disruptions and further downstream effects (Patz et al. 2003; Takaro, Knowlton, and Balmes 2013).

Climatic conditions impact the epidemiology of infectious disease and interact with behavioral, demographic, and socioeconomic factors, among others, to influence the incidence, emergence, and distribution of infectious diseases (Watts et al., "Shaping the Health of Nations," 2018). Despite an overall declining trend of infectious-disease-related mortality, it still accounts for 20 percent of the global burden of disease (Watts et al., "Shaping the Health of Nations," 2018). For instance, deaths from dengue fever were the highest in the Southeast Asia region, which includes Bangladesh, in 2016, and the overall trend is increasing based on data from 1990 to 2016 (figure 1.2).

For several climate-sensitive diseases, vectorial capacity is likely to be positively associated with increasing exposures to temperature and rainfall (Watts et al. 2019). These effects are most acutely felt by low- and middle-income countries across the world. Vectorial capacity is a measure of the average daily rate of subsequent cases in a susceptible population resulting from one infected case and is calculated using a formula including the vector-to-human transmission probability per bite, the human infectious period, the average vector biting rate, the extrinsic incubation period, and the daily survival period (Watts et al. 2019): in other words, "a measurement of the efficiency of vector-borne disease transmission" (Norris 2006). Climate suitability for climate-sensitive infectious diseases has increased globally (Watts et al. 2020). Vectorial capacity for the transmission of dengue from *Aedes aegypti* and *Aedes albopictus* mosquitoes has increased significantly worldwide by 3 percent and 6 percent respectively, compared with 1990 levels (figure 1.3). The number of cases of dengue fever recorded

FIGURE 1.1

Pathways by which climate change affects human health

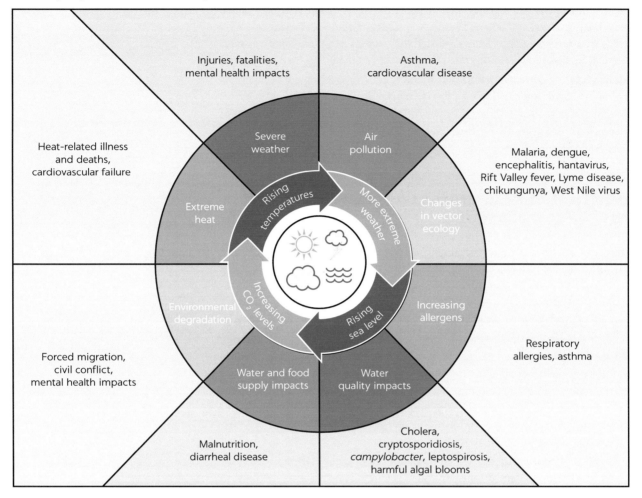

Source: World Bank Group and WHO 2018.
Note: CO₂ = carbon dioxide.

annually has doubled every decade since 1990, with 58.4 million apparent cases in 2013, accounting for more than 10,000 deaths (Watts et al. 2020). One of the potential factors that contributed to this increase is climate change. Other emerging and reemerging diseases, including yellow fever, chikungunya, Mayaro, and Zika viruses carried by *A. aegypti and A. albopictus* mosquitoes, are likely to be similarly responsive to the effects of climate change (Watts et al. 2020). Climate suitability for the Southeast Asia region for malaria has remained the same (figure 1.4), which implies that climate change is unlikely to alter the incidence and prevalence of the disease.

Based on this evidence, a few infectious diseases have been purposely grouped and selected for further analysis in this report. These include vector-borne diseases such as dengue, malaria, and chikungunya; waterborne diseases such as diarrhea and dysentery; and respiratory illnesses such as pneumonia and severe acute respiratory infection and associated symptoms. These are further detailed in "Persistent Illnesses" in chapter 4.

FIGURE 1.2

Global trends: Case mortality and mortality from selected causes as estimated by the global burden of disease, 1990–2016

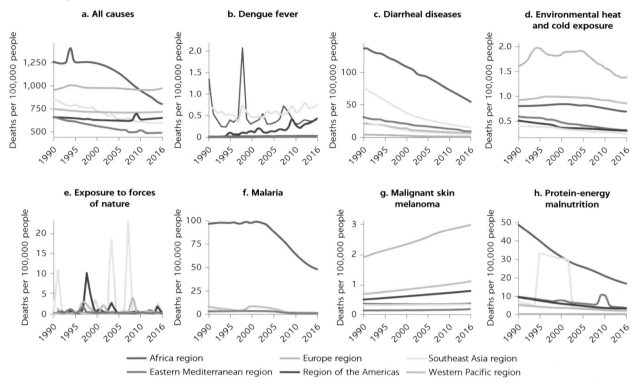

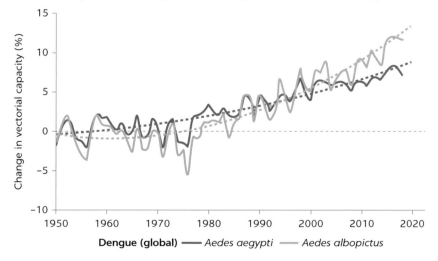

Africa region — Europe region — Southeast Asia region
Eastern Mediterranean region — Region of the Americas — Western Pacific region

Source: Watts et al., "Shaping the Health of Nations," 2018.
Note: The Southeast Asia region, including Bangladesh, is depicted by light green lines. For infectious diseases (malaria and diarrhea included in the figure), the mortality rate as measured by deaths per 100,000 people is declining over time in Southeast Asia, while for dengue fever, it has been increasing in recent years.

FIGURE 1.3

Vectorial capacity for dengue increasing over time across the globe

Dengue (global) — *Aedes aegypti* — *Aedes albopictus*

Source: Watts et al. 2020.
Note: Aedes aegypti and *Aedes albopictus* transmit dengue and other emerging and reemerging diseases including yellow fever, chikungunya, Mayaro, and Zika viruses. Vectorial capacity is defined as "a measurement of the efficiency of vector-borne disease transmission" (Norris 2006).

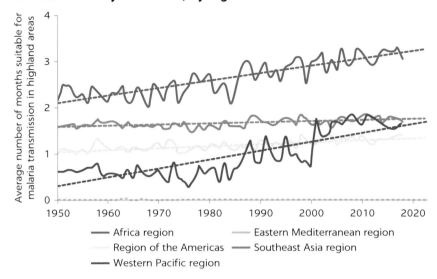

Source: Watts et al. 2020.
Note: Southeast Asia (including Bangladesh) is depicted by the orange line. Climate suitability for malaria transmission in Southeast Asia has remained comparable since 1950.

BANGLADESH'S VULNERABILITY TO CLIMATE CHANGE AND VARIABILITY

Bangladesh is a low-lying river delta with a long coastline of 711 kilometers and floodplains that occupy 80 percent of the country (Hasib and Chathoth 2016). The country experiences a multitude of natural disasters every year. Severe floods, cyclones, storms, tidal surges, and river erosion frequently cause loss of lives with devastating social and economic impacts. These extreme weather events are expected to be exacerbated by the effects of climate change (Rahman et al. 2019). The Government of Bangladesh's National Climate Vulnerability Assessment identified a number of climate-related hazards in 2018 that are critical for Bangladesh including increase in temperature and heat stress; increase in droughts; increase in rainfall intensity; increase in high river flows and flood risks; riverbank erosion; sea level rises and salinity intrusion; landslides; and increased intensity of cyclones, storm surges, and coastal flooding (Bangladesh, MoEFCC 2018).

In rural areas housing nearly 80 percent of the population, climate change has an immediate and direct effect on the health and well-being of millions of people who depend on natural resources for their basic livelihoods. The impacts of climate change are also increasingly felt in large cities that are exposed to various climate-induced hazards including variations in temperature, excessive and erratic rainfall, water logging, flooding, and heat and cold waves (Rabbani, Rahman, and Islam 2011). These hazards are exacerbated by high population density, poverty, rural-to-urban migration, illiteracy, and lack of public utilities and services. Rapid urbanization and a growing urban slum population are quickly changing the population dynamics in Bangladesh, and this has implications for climate-induced health risks (Mani and Wang 2014).

The country has the highest mortality rate in the world caused by natural disasters, with more than half a million people lost to disaster events since 1970.

The majority of these deaths have occurred during floods or cyclones (Nahar et al. 2014). More recently, Bangladesh was hit by two major cyclones: Sidr in 2007 and Aila in 2009. Cyclone Sidr killed 3,406 people, and more than 55,000 sustained physical injuries. Heavy rain and tidal waves owing to wind effects caused extensive physical destruction and damages to crops and livestock. After Cyclone Sidr, an assessment by the Government of Bangladesh found widespread outbreaks of diarrhea, dysentery, acute respiratory infection, and pneumonia. Children ages five years or younger were the most vulnerable (Kabir et al. 2016). Cyclone Aila hit the southern coastline of Bangladesh and partly damaged the Sundarbans. Along with outbreaks of diarrheal diseases was an acute scarcity of drinking water and food. With the number and intensity of such storms or cyclones projected to increase (see more details in the section "Definition of Key Terms Used in This Report" in chapter 3), climate change can reverse some of the significant gains Bangladesh has made in improving health-related outcomes, particularly in reducing child mortality, improving maternal health, and improving nutritional outcomes.

OBJECTIVES OF THE REPORT

Given the backdrop of evolving climatic conditions and their effect on health and the lack of representative evidence on this relationship, particularly for Bangladesh, the report aims to address the following set of questions:

What is the link between climate change and infectious diseases and what is the influence of climate variability on the mosquito life cycle, as it is one of the largest sources of vector-borne diseases in Bangladesh? The response to this question is derived through a systematic global and regional literature review to summarize existing evidence on the relationship of climate change or climate variability and infectious diseases and mental health. A summary of the literature related to environmental health that explores the relationship of weather patterns to mosquito life cycle is also presented.

How is the weather pattern evolving in Bangladesh? Leveraging historic weather data over the period 1976 to 2019 collected from the Bangladesh Meteorological Department, the report analyzes trends in temperature, rainfall, and humidity and constructs weather variables such as the heat index. Predicted changes to the weather for Bangladesh are also presented for an overview of the kinds of future weather events that are likely.

What is the relationship between climate variability, infectious diseases, and mental health? This question is answered by analyzing nationally representative household-level panel data from 3,600 households.[2] The document first reports on the prevailing rates of infectious diseases—further disaggregated by vector-borne, waterborne, and respiratory illnesses—and mental health issues, disaggregated by depression and anxiety disorders. Next, it establishes the relationship between weather outcomes and the likelihood of contracting any of the conditions.

The recommendations derived from this report are therefore context specific and driven by evidence. Specifically, recommendations for strengthening the health systems for dealing better with the effects of climate change, including the need for regular surveillance of diseases, are made for better mitigation.

The findings and recommendations are expected to assist practitioners and subject matter experts in national and global policy dialogue. It will similarly contribute to advancing the World Bank's corporate mandate relating to climate change. Policy dialogue will focus on supporting the government in further developing and implementing mitigation, adaptation, and resilience measures to tackle the effects of climate change.

APPROVAL AND CLEARANCE PROCESSES

Approval from the Bangladesh Medical Research and Council (BMRC) was obtained prior to commencement of the survey. All ethical protocols and standards of BMRC were adhered to during fieldwork. The following procedures were followed:

- Written informed consent of the interviewee was obtained.
- Names of respondents were not recorded; instead, a unique identity number was attached to the household. The privacy of information collected was ensured by keeping it anonymous (not attaching names of the respondents to the data).
- Data on nationality and religion were not collected.
- Respondents' personal information was not included in data files.
- Results were presented in aggregate form, without identifying any individual.

A concept note, outlining the overall objectives and structure of the report, was reviewed and approved by World Bank senior management. The note was also shared with the Climate Change and Health Promotion Unit (CCHPU) and the Institute of Epidemiology and Disease Control Research (IEDCR) of the Ministry of Health and Family Welfare of the Government of Bangladesh before finalizing the survey instrument and sample size.

The draft report was also shared with CCHPU and IEDCR before finalization. For quality assurance, the report was reviewed at an internal World Bank meeting, chaired by Mercy Tembon, country director for Bangladesh and Bhutan. The review was organized to discuss the methodology and findings and the potential implications of the conclusions and recommendations for Bangladesh. Based on detailed discussions during the internal review and extensive comments provided by the reviewers, the report was finalized. Reviewers included the following World Bank experts: Dhushyanth Raju (lead economist), Shiyong Wang (senior health specialist), Tamer Samah Rabie (lead health specialist), Anna Koziel (senior health specialist), Stephen Geoffrey Dorey (health specialist), and Muthukumara Mani (lead economist).

STRUCTURE OF THE REPORT

The introduction (chapter 1) provides the context, rationale, and objectives of the work, its contributions, and the intended audience. The distinction between the concepts of climate change and climate variability is made, and a theoretical framework is presented on the pathways of climate change's effect on health. Chapter 2 provides an overview of existing literature on the links between climate change or climate variability, infectious diseases, and mental health from a global perspective, and more specifically for Bangladesh through secondary analyses.

The report is arranged into four parts. Part I covers chapters 1 and chapter 2. Part II, covering chapters 3 and 4, presents findings from the household surveys conducted in 2019 and 2020. The former provides details of the survey, including definitions, sampling strategies, and methodologies used for analyses. The latter, chapter 4, presents the analyses from the household surveys.

Part III, which includes chapters 5 to 8, documents climate change patterns observed for Bangladesh, both historical and projected, and the patterns of selected infectious diseases.

Part IV, chapters 9 and 10, provides recommendations followed by concluding remarks. Supplementary details are provided in the appendixes.

NOTES

1. World Bank Climate Change Knowledge Portal (database), World Bank, Washington, DC. https://climateknowledgeportal.worldbank.org.
2. Data are representative of both urban and rural areas. The sampling outlined in chapter 3 distinguishes between two variations of urban area—major metropolises such as Dhaka and Chattogram cities, given their distinctive nature, and other urban areas. Data from the same households are collected, first during the monsoon and next during the dry seasons to account for seasonality.

REFERENCES

Balasubramanian, M. 2018. "Climate Change, Famine, and Low-Income Communities Challenge Sustainable Development Goals." *Lancet Planetary Health* 2 (10): e421–e422.

Bangladesh, MoEF (Ministry of Environment and Forests). 2008. *Bangladesh Climate Change Strategy and Action Plan.* Dhaka: MoEF. https://www.sdnbd.org/moef.pdf.

Bangladesh, MoEFCC (Ministry of Environment, Forests and Climate Change). 2018. *Nationwide Climate Vulnerability Assessment Bangladesh.* Final draft, November 2018. Dhaka: MoEFCC and German Agency for International Cooperation (GIZ).

Hasib, E., and P. Chathoth. 2016. "Health Impacts of Climate Change in Bangladesh: A Summary." *Current Urban Studies* 4 (1): 1–8. doi:10.4236/cus.2016.41001.

IPCC (International Panel on Climate Change). 2014. *Climate Change 2014: Synthesis Report. Contribution of Working Groups I, II and III to the Fifth Assessment Report of the Intergovernmental Panel on Climate Change.* Geneva: IPCC.

Kabir, R., H. T. A. Khan, E. Ball, and K. Caldwell. 2016. "Climate Change Impact: The Experience of the Coastal Areas of Bangladesh Affected by Cyclones Sidr and Aila." *Journal of Environmental and Public Health* 2016: 9654753. doi:10.1155/2016/9654753.

Lancet. 2019. "Health and Climate Change: Making the Link Matter." Editorial, November 13, 2019. *Lancet* 394 (10211): 1780. doi:10.1016/S0140-6736(19)32756-4.

Mani, M., and L. Wang. 2014. *Climate Change and Health Impacts: How Vulnerable Is Bangladesh and What Needs to Be Done.* Washington, DC: World Bank. https://openknowledge.worldbank.org/handle/10986/21820.

McMichael, T., H. Montgomery, and A. Costello. 2012. "Health Risks, Present and Future, from Global Climate Change." *BMJ* 344: e1359. doi:10.1136/bmj.e1359.

Nahar, N., Y. Blomstedt, B. Wu, I. Kandarina, L. Trisnantoro, and J. Kinsman. 2014. "Increasing the Provision of Mental Health Care for Vulnerable, Disaster-Affected People in Bangladesh." *BMC Public Health* 14 (July): 708. doi:10.1186/1471-2458-14-708.

Norris, Douglas E. 2006. "Malaria Entomology." Presentation. Johns Hopkins Bloomberg School of Public Health. https://www.glowm.com/pdf/M-lecture4.pdf.

Patz, G. A., J. P. McCarty, S. Hussein, U. Confalonieri, and N. D. Wet. 2003. "Climate Change and Infectious Diseases." In *Climate Change and Human Health: Risks and Responses*, edited by A. J. McMichael, D. H. Campbell-Lendrum, C. F. Corvalán, K. L. Ebi, A. K. Githeko, J. D. Scheraga, and A. Woodward, 103–32. Geneva: World Health Organization.

Rabbani, G., A. Rahman, and N. Islam. 2011. "Climate Change Implications for Dhaka City: A Need for Immediate Measures to Reduce Vulnerability." In *Resilient Cities: Cities and Adaptation to Climate Change*, vol 1., *Local Sustainability*, edited by K. Otto-Zimmerman. Dordrecht, ZUID-Holland: Springer. doi:10.1007/978-94-007-0785-6_52.

Rahman, M. M., S. Ahmad, A. S. Mahmud, M. Hassan-uz-Zaman, M. A. Nahian, A. Ahmed, Q. Nahar, and P. K. Streatfield. 2019. "Health Consequences of Climate Change in Bangladesh: An Overview of the Evidence, Knowledge Gaps and Challenges." *WIREs Climate Change* 10 (5): e601. doi:10.1002/wcc.601.

Takaro, T. K., K. Knowlton, and J. R. Balmes. 2013. Climate Change and Respiratory Health: Current Evidence and Knowledge Gaps. *Expert Review of Respiratory Medicine* 7 (4): 349–61.

Watts, N., M. Amann, N. Arnell, S. Ayeb-Karlsson, L. Beagley, K. Belesova, M. Boykoff, et al. 2020. *Responding to Convergence Crises*. The 2020 report of the *Lancet* Countdown on health and climate change. *Lancet* 397 (10269): 129–70. doi:10.1016/S0140-6736(20)32290-X.

Watts, N., M. Amann, N. Arnell, S. Ayeb-Karlsson, K. Belesova, H. Berry, T. Bouley, et al. 2018. *Shaping the Health of Nations for Centuries to Come*. The 2018 report of the *Lancet* Countdown on health and climate change, *Lancet* 392 (10163): 2479–514. doi:10.1016/S0140-6736(18)32594-7.

Watts, N., M. Amann, N. Arnell, S. Ayeb-Karlsson, K. Belesova, M. Boykoff, P. Byass, et al. 2019. *Ensuring That the Health of a Child Born Today Is Not Defined by a Changing Climate*. The 2019 report of the *Lancet* Countdown on health and climate change, *Lancet* 394 (10211): 1836–78. doi:10.1016/S0140-6736(19)32596-6.

Watts, N., M. Amann, S. Ayeb-Karlsson, K. Belesova, T. Bouley, M. Boykoff, P. Byass, et al. 2018. *From 25 Years of Inaction to a Global Transformation for Public Health*. The 2018 report of the *Lancet* Countdown on health and climate change, *Lancet* 391 (10120): 581–630. doi:10.1016/S0140-6736(17)32464-9.

WHO (World Health Organization). 2003. *Climate Change and Human Health: Risk and Responses*. Geneva: WHO. https://www.who.int/globalchange/publications/climchange.pdf.

WHO (World Health Organization). 2016. "An Estimated 12.6 Million Deaths Each Year Are Attributable to Unhealthy Environments." Press Release, March 6, 2016. http://www.who.int/mediacentre/news/releases/2016/deaths-attributable-to-unhealthy-environments/en.

WHO and UNFCCC (World Health Organization and United Nations Framework Convention on Climate Change). 2015. *Climate and Health Country Profile 2015: Bangladesh*. Geneva: WHO.

World Bank. 2012. *Turn Down the Heat: Why a 4°C Warmer World Must Be Avoided*. Washington, DC: World Bank. https://openknowledge.worldbank.org/handle/10986/11860.

World Bank Group and WHO. 2018. *Methodological Guidance: Climate Change and Health Diagnostic, A Country-Based Approach for Assessing Risks and Investing in Climate-Smart Health Systems*. Investing in Climate Change and Health Series. Washington, DC: World Bank Group; Geneva, WHO.

2 Overview of Evidence Gathered

INTRODUCTION

This chapter thematically summarizes the relevant literature on the effects of climate change or climate variability on infectious diseases and mental health. Approximately 180 reports and papers were reviewed during the process. The overview first presents a global context of the relevant topics followed by a focus on issues specifically pertinent to Bangladesh. Appendix A includes details.

INFECTIOUS DISEASES

Climate change, which includes alterations in one or more variables such as temperature, rainfall, sea-level elevation, wind, and duration of sunlight, affects many climate-sensitive infectious diseases through the survival, reproduction, or distribution of disease pathogens and hosts as well as the availability and means of their transmission environment (Wu et al. 2016). Human behavior such as crowding and displacement amplify risks of infection (McMichael, Montgomery, and Costello 2012). An agent (pathogen), a vector (host), and favorable transmission environment are three components essential for the spread of an infectious disease (Wu et al. 2016). A limited range of climatic conditions facilitates the climate envelope within which each infective agent or vector species can survive and reproduce (Patz et al. 2003).

There is sufficient observational evidence on the effects of meteorological factors on the incidence of vector-borne, waterborne, airborne, and food-borne diseases. A more contemporary concern is the extent to which changes in disease patterns will occur under the conditions of global climate change (Patz et al. 2003). The correlation between meteorological factors and the components of transmission cycles such as parasite development rates, vector biting, and survival rates or the observed geographical distribution of diseases have been used to generate predictive models (Campbell-Lendrum et al. 2015). These models link projections of future scenarios of climate change with other determinants, such as gross domestic product—as a measure of socioeconomic and technological development—and urbanization. However, because of uncertainties in climate projections and future development trends, as well as the compounding effect of natural climate variability over short to medium timescales—from years to within two decades—the models are highly approximate and are able to comment only on broad trends.

In Bangladesh, very few studies have explored the relationship between environmental variables and infectious diseases. Temperature and precipitation changes have been found to impact the dynamics of vector-borne diseases such as malaria, dengue, visceral leishmaniasis—commonly known as kala-azar—cholera, and diarrheal diseases (Rahman et al. 2019; Banu et al. 2014; Hossain, Noiri, and Moji 2011; Reid et al. 2012; Hashizume et al. 2007). Although the country has made progress in controlling communicable diseases in recent years, dengue cases have surged, as have chikungunya and Zika cases more recently, causing major threats to the health of the population. Higher temperatures are expected to increase transmission and spread of vector-borne diseases by increasing mosquito density in some areas and increasing replication rate and bite frequency of mosquitoes (Costello et al. 2009). This will, in turn, increase the incidence of malaria, dengue, and tick-borne encephalitis.

MENTAL HEALTH

Global evidence on the effects of climate change or climate variability on mental health is limited but is steadily increasing. Extreme weather events brought on by climate change have been attributed as one of the triggers of a host of mental health issues (Berry, Bowen, and Kjellstrom 2010). These include major depressive disorders and other forms of depression, anxiety, post-traumatic stress disorder (PTSD), grief and bereavement, survivor guilt, recovery fatigue, substance abuse, suicidality, and vicarious trauma in first responders (Berry 2009; Berry, Bowen, and Kjellstrom 2010; Berry et al. 2008; Bourque and Willox 2014; Clayton et al. 2017; Coyle and Susteren 2012; Doherty and Clayton 2011; Swim et al. 2009; Weissbecker 2011; Willox et al., "The Land Enriches," 2013; Willox et al., "Climate Change," 2013; Willox et al. 2015). Most of the evidence on the topic, however, pertains to high-income countries. Based on the limited insights available from low- and middle-income countries, mental health issues are likely to be aggravated, given the existing vulnerabilities and limited capacity to address mental health issues (WHO 2009).

Natural disasters and environmental degradation because of climate change or climate variability or of both together are known risk factors that can affect the psychological health of vulnerable populations in Bangladesh, especially those living in coastal areas—although this has not been documented well in the local context. The majority of the people living in Bangladesh's coastal areas are low-income agricultural workers, many of whom are landless and are relatively asset poor (Bangladesh, MoEF 2008; Paul 2009). They are frequently affected by natural disasters but have insufficient resources to protect themselves, to adequately rebuild their lives after the event, or to access the medical services when needed (Nahar et al. 2014). The initial response is to ensure that survivors receive the basic necessities to sustain life such as shelter, food, safe water, and sanitation. However, after this acute, emergency phase, many of the affected populations or climate refugees are left with some level of psychological or mental health problems. Some of these include PTSD, depressive symptoms or major depressive disorders, anxiety or generalized anxiety disorders, as well as more general mental health problems such as sleep disruption, substance abuse, and aggression (Norris 2005; Paul, Rahman, and Rakshit 2011).

CLIMATE VARIABILITY AND MOSQUITOES

Dengue is one of the most important mosquito-borne diseases affected by climate variability, and it continues to spread throughout the tropical and subtropical regions globally (Costa et al. 2010). Dengue, chikungunya, and Zika virus are spread by the same mosquito species, *Aedes aegypti* (Lowe et al. 2017). Figure 2.1 presents the pathways by which dengue transmission cycles are altered by weather variables and other factors.

Ebi and Nealon (2016) summarize the life cycle of a mosquito: female mosquitoes lay eggs on the side of water-holding containers while humans provide the blood meals necessary for egg development. These female mosquitoes usually rest in cool and dark places and generally bite humans indoors. After flooding or rain, the eggs hatch into larvae. In a week or so, under favorable environmental conditions, the larvae transform into pupae and evolve into adult mosquitoes. With respect to the viruses spread by these mosquitoes, it takes between 5 and 33 days, with a mean of 15 days, at 25°C for the virus to multiply, mature, and travel to the salivary glands of the mosquito before the insect can start transmitting the virus by biting a person.

The variability in climatic conditions—temperature, precipitation or rainfall, and humidity—because of climate change will affect the biology of mosquito vectors as well as the risk of disease transmission (Costa et al. 2010). Colón-González, Lake, and Hunter (2013) conclude that dengue transmission rapidly increases

FIGURE 2.1

WHO and World Meteorological Organization framework on the interaction of meteorological and other determinants of dengue transmission cycles and clinical diseases

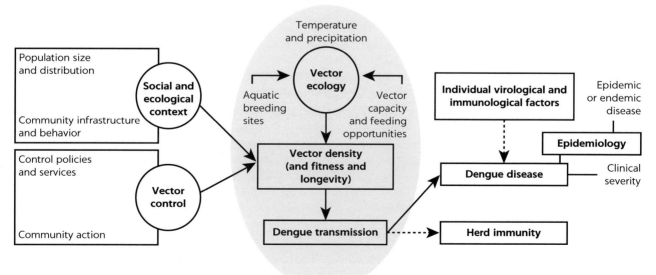

Source: Ebi and Nealon 2016.

FIGURE 2.2

Relationship between incidence of dengue and minimum temperature, maximum temperature, and rainfall

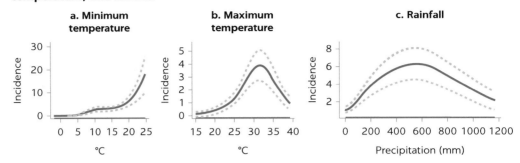

Source: Colón-González, Lake, and Hunter 2013.
Note: Data are from laboratory-confirmed dengue cases in Mexico from 1985 to 2007. Solid lines indicate the average expected number of dengue cases per 100,000 people per month; dashed lines indicate the estimated 95% confidence intervals.

when the minimum temperature increases above 18°C, based on data from Mexico on laboratory-confirmed dengue cases from 1985 to 2007 and weather data—monthly averages for minimum temperature, maximum temperature, and rainfall (figure 2.2). They conclude that the minimum temperature has the biggest impact on dengue—with zero risk below 5°C and a rapid increased risk when the average minimum temperature is above 18°C. The maximum temperature also influences dengue independently from the minimum temperature; the authors found that dengue cases increase in the range of 25°C to 35°C with a peak at 32°C. At temperatures above 32°C the risk of dengue decreases, with adult mosquitoes dying at temperatures above 35°C. With respect to rainfall, dengue cases increase in the range of 200 to 800 millimeters of rainfall, with a peak at 550–650 millimeters. The authors also found higher incidence of dengue in the wet season from May to October for Mexico. Zhang et al. (2019) conclude that periods of increased temperatures can cause occurrence of dengue epidemics.

REFERENCES

Bangladesh, MoEF (Ministry of Environment and Forests). 2008. *Bangladesh Climate Change Strategy and Action Plan.* Dhaka: MoEF. https://www.sdnbd.org/moef.pdf.

Banu, S., W. Hu, Y. Guo, C. Hurst, and S. Tong. 2014. "Projecting the Impact of Climate Change on Dengue Transmission in Dhaka, Bangladesh." *Environment International* 63: 137–42.

Berry, H. 2009. "Pearl in the Oyster: Climate Change as a Mental Health Opportunity." *Australian Psychiatry* 17 (6): 453–56.

Berry, H. L., K. Bowen, and T. Kjellstrom. 2010. "Climate Change and Mental Health: A Causal Pathways Framework." *International Journal of Public Health* 55 (2): 123–32.

Berry, H. L., B. J. Kelly, I. C. Hanigan, J. H. Coates, A. J. McMichael, J. A. Welsh, and T. Kjellstrom. 2008. *Rural Mental Health Impacts of Climate Change.* Canberra: Australian National University.

Bourque, F., and A. C. Willox. 2014. "Climate Change: The Next Challenge for Public Mental Health?" *International Review of Psychiatry* 26 (4): 415–22.

Campbell-Lendrum, D., L. Manga, M. Bagayoko, and J. Sommerfeld. 2015. "Climate Change and Vector-Borne Diseases: What Are the Implications for Public Health Research and Policy?" *Philosophical Transactions of the Royal Society B* 370: 20130552. doi:10.1098/rstb.2013.0552.

Clayton, S., C. Manning, K. Krygsman, and M. Speiser. 2017. *Mental Health and Our Changing Climate: Impacts, Implications, and Guidance.* Washington, DC: American Psychological Association and ecoAmerica.

Colón-González, F. J., I. Lake, and P. R. Hunter. 2013. "The Effects of Weather and Climate Change on Dengue." *PLoS Neglected Tropical Diseases* 7 (11): e2503. doi:10.1371/journal .pntd.0002503.

Costa, E. A. P. de A., E. M. de M. Santos, J. C. Correia, and C. M. R. de Albuquerque. 2010. "Impact of Small Variations in Temperature and Humidity on the Reproductive Activity and Survival of *Aedes aegypti* (Diptera, Culcidae)." *Revista Brasileira de Entomologia* 54 (3): 488–93.

Costello, A., M. Abbas, A. Allen, S. Ball, S. Bell, R. Bellamy, S. Friel, et al. 2009. "Managing the Health Effects of Climate Change." *Lancet* 373: 1693–1733.

Coyle, K. J., and L. Van Susteren. 2012. *The Psychological Effects of Global Warming on the United States: And Why the US Mental Health Care System Is Not Adequately Prepared.* Washington, DC: National Wildlife Federation. https://www.nwf.org/~/media/PDFs/Global-Warming /Reports/Psych_effects_Climate_Change_Ex_Sum_3_23.ashx.

Doherty, T. J., and S. Clayton. 2011. "The Psychological Impacts of Global Climate Change." *American Psychologist* 66 (4): 265.

Ebi, K., and Nealon, J. 2016. "Dengue in a Changing Climate." *Environmental Research* 151 (2016): 115–23.

Hashizume, M., B. Armstrong, S. Hajat, Y. Wagatsuma, A. S. Faruque, T. Hayashi, and D. A. Sack. 2007. "Association between Climate Variability and Hospital Visits for Non-Cholera Diarrhea in Bangladesh: Effects and Vulnerable Groups." *International Journal of Epidemiology* 36 (5): 1030–37. doi:10.1093/ije/dym148.

Hossain, M., E. Noiri, and K. Moji. 2011. "Climate Change and Kala-azar." In *Kala-azar in South Asia*, edited by E. Noiri and T. K. Jha, 127–37. Dordrecht, Netherlands: Springer. doi:10.1007/978-94-007-0277-6_12.

Lowe, R., A. M. Stewart-Ibarra, D. Petrova, M. Garcia-Díez, M. J. Borbor-Cordova, R. Mejía, M. Regato, and X. Radó. 2017. "Climate Services for Health: Predicting the Evolution of the 2016 Dengue Season in Machala, Ecuador." *Lancet Planet Health* 1: e142–e151.

McMichael, T., H. Montgomery, and A. Costello. 2012. "Health Risks, Present and Future, from Global Climate Change." *BMJ* 344: e1359. doi:10.1136/bmj.e1359.

Nahar, N., Y. Blomstedt, B. Wu, I. Kandarina, L. Trisnantoro, and J. Kinsman. 2014. "Increasing the Provision of Mental Health Care for Vulnerable, Disaster-Affected People in Bangladesh." *BMC Public Health* 14 (July): 708. doi:10.1186/1471-2458-14-708.

Norris, F. H. 2005. *Psychosocial Consequences of Natural Disasters in Developing Countries: What Does Past Research Tell Us about the Potential Effects of the 2004 Tsunami?* Washington, DC: National Center for PTSD, US Department of Veterans Affairs.

Patz, G. A., J. P. McCarty, S. Hussein, U. Confalonieri, and N. D. Wet. 2003. "Climate Change and Infectious Diseases." In *Climate Change and Human Health: Risks and Responses*, edited by A. J. McMichael, D. H. Campbell-Lendrum, C. F. Corvalán, K. L. Ebi, A. K. Githeko, J. D. Scheraga, and A. Woodward, 103–32. Geneva: World Health Organization.

Paul, B. K. 2009. "Why Relatively Fewer People Died? The Case of Bangladesh's Cyclone Sidr." *Natural Hazards* 50: 289–304. doi:10.1007/s11069-008-9340-5.

Paul, B. K., M. K. Rahman, and B. C. Rakshit. 2011. "Post-Cyclone Sidr Illness Patterns in Coastal Bangladesh: An Empirical Study." *Natural Hazards* 56: 841–52. doi:10.1007/s11069 -010-9595-5.

Rahman, M. M., S. Ahmad, A. S. Mahmud, M. Hassan-uz-Zaman, M. A. Nahian, A. Ahmed, Q. Nahar, and P. K. Streatfield. 2019. "Health Consequences of Climate Change in Bangladesh: An Overview of the Evidence, Knowledge Gaps and Challenges." *WIREs Climate Change* 10 (5): e601. doi:10.1002/wcc.601.

Reid, H. L., U. Haque, S. Roy, N. Islam, and A. C. Clements. 2012. "Characterizing the Spatial and Temporal Variation of Malaria Incidence in Bangladesh, 2007." *Malaria Journal* 11 (170): 170. doi:10.1186/1475-2875-11-170.

Swim, J., S. Clayton, T. Doherty, R. Gifford, G. Howard, J. Reser, P. Stern, and E. Weber. 2009. *Psychology and Global Climate Change: Addressing a Multi-Faceted Phenomenon and Set of Challenges*. Washington, DC: American Psychological Association.

Weissbecker, I. 2011. *Climate Change and Human Well-Being: Global Challenges and Opportunities*. Berlin: Springer.

WHO (World Health Organization). 2009. *Mental Health Systems in Selected Low- and Middle-Income Countries: A WHO-AIMS Cross-National Analysis*. Geneva: WHO. https://www.who.int/mental_health/evidence/WHO-AIMS/en.

Willox, A. C., S. L. Harper, V. L. Edge, K. Landman, K. Houle, and J. D. Ford. 2013. "The Land Enriches the Soul: On Climatic and Environmental Change, Affect, and Emotional Health and Well-Being in Rigolet, Nunatsiavut, Canada." *Emotion, Space, and Society* 6: 14–24.

Willox, A. C., S. L. Harper, J. D. Ford, V. L. Edge, K. Landman, K. Houle, S. Blake, and C. Wolfrey. 2013. "Climate Change and Mental Health: An Exploratory Case Study from Rigolet, Nunatsiavut, Canada." *Climatic Change* 121 (2): 255–70.

Willox, A. C., E. Stephenson, J. Allen, F. Bourque, A. Drossos, S. Elgarøy, M. J. Kral, et al. 2015. "Examining Relationships between Climate Change and Mental Health in the Circumpolar North." *Regional Environmental Change* 15. doi:10.1007/s10113-014-0630-z.

Wu, X., Y. Lu, S. Zhou, L. Chen, and B. Xu. 2016. "Impact of Climate Change on Human Infectious Diseases: Empirical Evidence and Human Adaptation." *Environment International* 86: 14–23. doi:10.1016/j.envint.2015.09.007.

Zhang, Q., Y. Chen, Y. Fu, T. Liu, Q. Zhang, P. Guo, and W. Ma. 2019. "Epidemiology of Dengue and the Effect of Seasonal Climate Variation on Its Dynamics: A Spatio-Temporal Descriptive Analysis in the Chao-Shan Area on China's Southeastern Coast." *BMJ Open* 9: e024197.

Analysis of Primary Data Collected through a Survey

3 Data and Methods

INTRODUCTION

Several sources of data have been used in this report. These include primary panel data collected in two rounds—August and September 2019 and January and February 2020—localized weather data from the Bangladesh Meteorological Department (BMD) covering conditions from the two months preceding each survey, and secondary analysis of data available from various sources.

HOUSEHOLD PANEL DATA

The first round canvassed 3,610 households comprising 15,383 individuals, between August and September 2019 immediately past the peak of the monsoon season. The follow-up round collected the same information from the same households between January and February 2020 during the dry season. The second round canvassed 3,480 households comprising 14,474 individuals, with an attrition rate of 3 percent. The timings of the two rounds were deliberately chosen to identify seasonality and variations in the outcomes of interest. Households were tracked for any change of their residence between the survey periods. The sample is representative of urban and rural areas. The sampling design allows for assessing heterogeneity across urban areas, such as major city centers and other urban areas.

Survey design and sampling strategy

The survey was designed using a two-stage stratified random sampling. The primary sampling units (PSUs) in the first stage were selected using probability-proportional-to-size methods utilizing the 2011 population and housing census. Map 3.1 presents the distribution of PSUs across the country. The 150 PSUs of the first stage were selected based on three strata to account for levels of congestion. The first stratum represents rural areas comprising 90 PSUs, the second represents Dhaka and Chattogram city corporation areas

MAP 3.1

Sample primary sampling units by enumeration areas

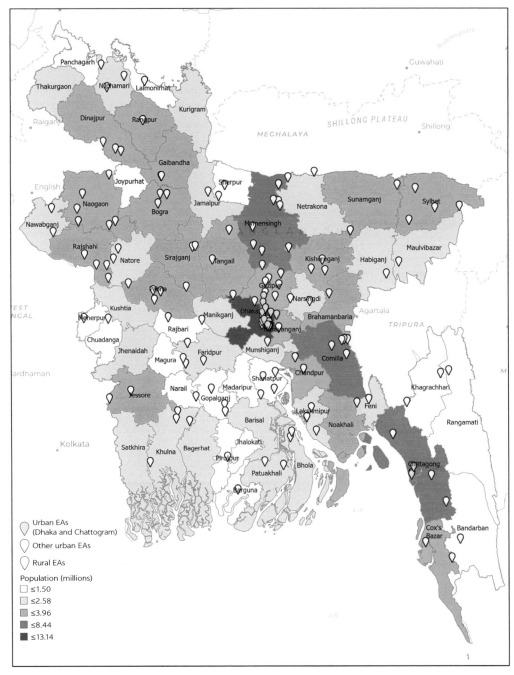

Source: Original map for this publication, based on data from Esri, HERE Technologies, Garmin, the Food and Agriculture Organization of the United Nations, the National Oceanic and Atmospheric Administration of the United States, and the United States Geological Survey.
Note: EA = enumeration area.

comprising 24 PSUs, and the third stratum consists of 36 PSUs representing all urban areas (table 3.1).

The second stage of the selection process in each of the enumeration areas (EAs) began with a listing exercise. For very large EAs, a smaller section was delineated for the listing. The second level of stratification is defined as

TABLE 3.1 **Household sample, 2019–20**

		HOUSEHOLDS	
	CLUSTERS	ROUND 1 (AUG–SEP 2019)	ROUND 2 (JAN–FEB 2020)
Rural areas	90	2,164	2,125
Dhaka and Chattogram cities	24	580	517
All urban areas (including Dhaka and Chattogram)	36	866	838
National	**150**	**3,610**	**3,480**

Source: Original table for this publication.

(1) households with women and children under five years of age, (2) households with an elderly population (age 65 years or more), and (3) households with mixed demographics. Households were randomly selected from each stratum with the predetermined ratio of 16:2:2. For EAs where the ratio was unable to be attained because of the absence of households in certain strata, replacements were obtained from the first category to arrive at a final number of 20 observations per EA.

Sampling weights are calculated in two stages. For the first stage, the probability of selection of the sample was calculated separately for each stage and EA using the following specification in equation 3.1:

$$P_{ij}^1 = n_j \frac{M_{ij}}{\sum M_{ij}},$$ (3.1)

where the probability of the ith EA being selected from the jth stratum is presented by P_{ij}^1. n_j represents the number of selected EAs within each stratum. The number of households in the ith EA in the jth stratum is represented by M_{ij} while $\sum M_{ij}$ represents the total number of households in the stratum.

The probability of the second stage of the selection process is estimated using the following specification in equation 3.2,

$$P_{ij}^2 = \frac{h_{tij}}{H_{tij}},$$ (3.2)

where the probability of a household being selected for the sample is represented by P_{ij}^2. h_{tij} are the number of households of type t in EA i in stratum j selected to be surveyed from a total number of households (H_{tij}) within each EA and the particular category.

The overall likelihood of a particular household being selected from a particular stratum is therefore represented as the product of the two aforementioned probabilities, calculated as shown in equation 3.3:

$$P_{ij} = P_{ij}^1 \times P_{ij}^2.$$ (3.3)

The weight is subsequently constructed as the inverse of the likelihood of a particular household being selected ($1/P_{ij}$).

The first stage involves the calculation of probability of each cluster being sampled within each stratum. The probability of selection of households within each stratum is calculated from the total listed households from each of the strata in each EA. The overall likelihood of a household being selected from each stratum is, therefore, calculated as the product of the two probabilities. The household weights are subsequently constructed as the inverse of the likelihood of a household being selected.

Data

A structured questionnaire, directed toward the primary female member of the household, was used to collect information on an array of issues. The cascading questions first inquired as to whether any member of the household fell ill followed by whether they had visited a doctor for the illnesses and had received a medical diagnosis. The subsequent set of questions collected detailed symptoms of illnesses in the event they had not acquired a medical diagnosis. This report considers three sets of primary outcomes—infectious diseases, persistent or chronic illnesses, and mental health. During the survey, respondents were asked about their morbidities. The types of diseases were grouped as infectious diseases and persistent illnesses (as defined in "Persistent Illnesses" in chapter 4).

Details were collected on health-care-seeking behavior, conditional on reporting either infectious or persistent illnesses. Questions were asked related to whether health care was sought, the duration between the illness and care, details on the type and location of the care provider, mode and duration of travel time, and direct (consultation, medication, diagnostics) and indirect (transportation) costs of care.

Data on the status of common mental disorders, namely, depression and anxiety, were collected. Depression is measured using the Patient Health Questionnaire-9 (PHQ-9), a commonly used depression screening instrument, comprising nine items on a four-point Likert response scale. The scale has demonstrated good sensitivity and specificity in identifying depression in both clinical and nonclinical settings (Levis, Benedetti, and Thombs 2019; Spitzer, Kroenke, and Williams 1999). The PHQ-9 has been validated for use in Bengali-speaking populations (Chowdhury, Ghosh, and Sanyal 2004) and has been widely used in Bangladesh (Arafat et al. 2018; Islam et al. 2020; Islam, Rawal, and Niessen 2015; Mamun et al. 2019; Moonajilin, Rahman, and Islam 2020; Roy et al. 2012). For anxiety, the Generalized Anxiety Disorder-7 (GAD-7), a seven-item, four-point, Likert-style anxiety screening scale was used. The GAD-7 has good reliability and validity for measuring generalized anxiety disorder (Spitzer et al. 2006) and has also been used previously in clinical and nonclinical research settings in Bangladesh (Hossain et al. 2019; Islam et al. 2020; Moonajilin, Rahman, and Islam 2020). Informed by best practice standards in existing literature (Levis, Benedetti, and Thombs 2019; Manea, Gilbody, and McMillan 2015; Spitzer et al. 2006), a clinical cutoff score of 10 has been used in this report to establish the presence of depression and anxiety for both PHQ-9 and GAD-7 scores.

Background information was collected also on a host of related issues. A roster held the amassed information on the households' demographic composition, and disability status was compiled using the short Washington Group Questions.[1] A composite socioeconomic index, accounting for unique urban and rural characteristics, was created from physical household attributes—wall and roof material, area per capita, and availability of a separate kitchen—and a roster of durable assets was created using methods outlined in a report by the National Institute of Population Research and Training (Bangladesh NIPORT, Mita and Associates, and ICF 2016). Quintiles of the continuous index are used in the rest of the report. A similar index for water, sanitation, and hygiene (WASH) was constructed using WASH attributes of the household, such as accessibility to sanitary facilities, type of water used, and hygiene-oriented behavior such as hand washing, among others. The WASH index is categorized as low, medium, and high, with low indicating access to fewer WASH facilities, low use, and so on. A community questionnaire was administered with the goal of assessing community-level characteristics such as relative poverty scenario, connectivity, and cleanliness, among others. A connectivity index has been constructed using community-level outcomes such as its size, construction material of the main road, average hours of electricity availability, whether a community-level system is available for collecting refuse, experience of waterlogging the previous year, and access to community-level services such as transport, post office, schools, and banks. Community-level information was collected from key respondents deemed to have a good understanding of the issues surrounding the locality, typically the community leaders. An array of information was collected related to socioeconomic status of average households; issues faced such as waterlogging or road conditions; access to basic services such as schools, police, post offices, markets, financial services; and so forth.

Respondent profile

Figure 3.1 shows the projected national population distribution from the sample and table B.1 (appendix B) provides details of demographic profiles of the sample. The sample is equally distributed between the sexes and remains consistent across the geographic strata. The average age is approximately 28 years. Almost half the population on average is married. The approximate educational attainment is 4.9 years of schooling with the population in urban areas better educated than their rural peers. A high proportion of the household heads are male at 92 percent, and the proportion is 4 percentage points lower in the cities of Dhaka and Chattogram than in rural areas. Similarly, the average age of the household heads are nearly twice the national average and they are better educated in urban than in rural areas.

Table B.2 (appendix B) outlines the socioeconomic conditions of the sampled households at baseline. The largest proportion of households at 24 percent falls in the lowest quintile of the socioeconomic index while the smallest at 16 percent is in the highest quintile. The poorest are most heavily represented from rural areas at 28 percent than all urban, and the cities of Dhaka and Chattogram are at 10 and 6 percent, respectively. Inversely, the richest reside in Dhaka and Chattogram and all urban areas—45 and 35 percent, respectively. This is reflected in other household characteristics. Compared to 22 percent in rural areas, none of the households in Dhaka and Chattogram and only 4 percent of households in all urban areas have mud or straw as the primary building material for walls. Ninety percent of the houses use tin as roof material in rural areas compared to 48 percent in Dhaka and Chattogram cities and 68 percent in all urban areas. Access to electricity is nearly universal across the urban space compared to 86 percent in rural areas. Similarly, 97 percent of the households in Dhaka and Chattogram cities use clean stoves in comparison to half of those in all urban areas and only 10 percent in rural areas. More household members share their rooms in Dhaka and Chattogram than they do in all urban or rural areas—0.46 rooms per capita in Dhaka and Chattogram cities compared to 0.54 rooms in the rest of the country.

FIGURE 3.1
Projected population distribution from sampled households, 2019–20

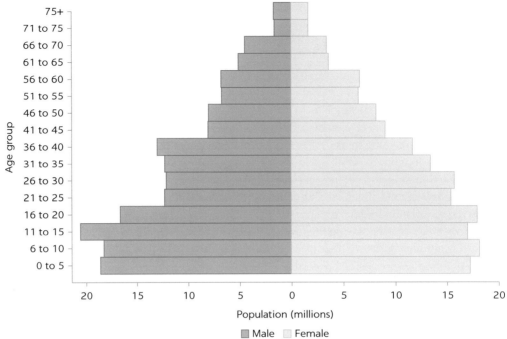

Source: Original figure for this publication.

Community profile

Table 3.2 presents community profiles included in the household survey. The primary road to access the area is paved for 97 percent of urban areas compared to 76 percent in rural areas. While all locations in the survey have access to electricity, urban spaces enjoy the service two hours longer on average than rural ones, due to power outages in the latter. Compared to almost 60 percent in Dhaka and Chattogram cities, only 16 percent of communities in the rural areas have experienced waterlogging in the year previous to this survey. The presence of stagnant water bodies is 98 percent in rural areas compared to 72 percent in all urban areas and 22 percent in Dhaka and Chattogram cities. All communities in Dhaka and Chattogram cities, half in all urban areas, and only a quarter in rural areas have a designated space to dispose of household refuse.

Figure 3.2 shows access to services by communities where the surveys were implemented, disaggregated by urban and rural areas. While access to improved water sources and primary schools is almost 100 percent, approximately 80 percent of the locals have access to secondary schools or markets. While access to post offices are widely reported in rural areas, the proportion of that access in the urban space is nearly half. The proportion of communities with access to essential services such as the police or fire services varies considerably between urban and rural areas. A fifth of the communities reported housing a police station in the locality in rural areas compared with 46 percent in urban areas. Only 3 percent of the rural communities included in the survey have a fire station in their village compared with 25 percent in the urban spaces.

TABLE 3.2 **Community profile, 2019–20**

mean

VARIABLES	NATIONAL	URBAN		RURAL
		ALL	DHK AND CTG	
Main access road is paved (%)	81	97	100	76
Hours of electricity available	20.04	22.26	22.89	19.47
Community experienced waterlogging previous year (%)	20	35	59	16
Community contains a stagnant water body (%)	93	72	22	98
Garbage generally disposed of in designated place (%)	31	53	100	25
Number of households in community	538	811	1391	468
N = 150				

Source: Original table for this publication.
Note: CTG = Chattogram; DHK = Dhaka.

FIGURE 3.2

Access to services, 2019–20

percent

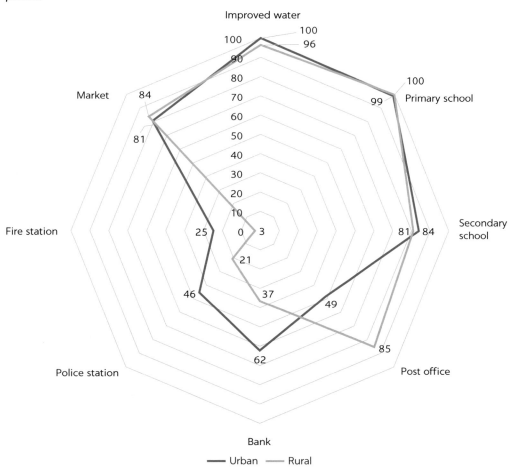

Source: Original figure for this publication.

Lastly, existence of financial services such as banks are 25 percentage points higher in urban areas than in rural areas.

Calculation of correlation coefficients

Correlation coefficients presented for infectious diseases and mental health in chapter 5 are estimated using the weighted linear specification as given in equation 3.4:

$$Y_{itz} = \alpha + \beta_1 \theta_j + \beta_2 \theta_j * t_t + t_t + \emptyset_z + \varepsilon_{itz} \qquad (3.4)$$

where Y_{itz} represens the physical or mental health outcome for the ith individual at time t living in the zth PSU. β_2 represents the coefficient of the interaction term identifying the seasonal differentials. PSU-level fixed effects are represented by $\emptyset_z$ while t_t reflects time trends. The decision to use PSU levels as opposed to exploiting the panel nature of the data through individual- or household-level fixed effects in the models is primarily because of the absence of positive illness outcomes in the sample. Approximately 5 percent of the total sample report an illness during the two surveys. Similarly, the issue precludes further assessment of heterogeneity through interaction of demographic or socioeconomic and weather variables. The use of PSU-level fixed effects captures and controls for locational trends over the two seasons in the models. The idiosyncratic error term is represented by ε_{itz}. Interpretation of binary explanatory outcomes are in percentage changes (for example, 0.03 = 3 percentage points) and in percent for linear outcomes (0.03 = 3 percent).

WEATHER DATA

Weather data were collected from the BMD between 1976 and 2019. The BMD collects data from 43 weather stations across the country. Map 3.2 shows the distribution of these stations, relative to population density. These 43 stations have been set up over time, which is why data for some of the earlier years, between 1976 and 1990, are from fewer weather stations.

DEFINITION OF KEY TERMS USED IN THIS REPORT

Climate is defined as "the description in terms of the mean and variability of relevant atmospheric variables such as temperature, precipitation and wind. Climate can thus be viewed as a synthesis or aggregate of weather" (Fouque and Reeder 2019, 2).

Climate variability refers "to the day-to-day change in meteorological parameters including temperature, precipitation, humidity, and winds. Extreme weather events are significant deviations of meteorological variables, such as floods caused by excessive rainfall, droughts, storm surges, and heat waves (extreme temperature). Climate variability refers to short-term changes in the

MAP 3.2

Weather station locations in Bangladesh

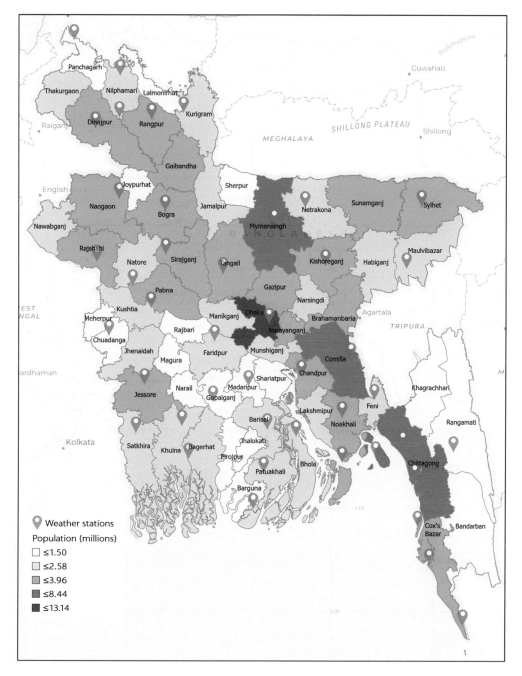

Source: Original map for this publication, based on data from Esri, HERE Technologies, Garmin, the Food and Agriculture Organization of the United Nations, the National Oceanic and Atmospheric Administration of the United States, and the United States Geological Survey.

average meteorological conditions over a time scale, such as a month, a season, or a year" (Mani and Wang 2014, 15).

Climate change refers to "changes in average meteorological conditions and seasonal patterns over a much longer time horizon, often over 50 or 100 years" (Mani and Wang 2014, 15).

Infectious diseases are defined as any seasonal disease experienced in the 30 days preceding the survey. Infectious diseases include diseases diagnosed by a medical professional as well as symptoms that the respondents experienced for which they did not seek any medical help. For the purpose of analyzing, these symptoms were tagged to specific diseases. Table 3.3 provides details of the categorization used for seasonal illnesses. These diseases are infectious and likely to be climate sensitive.

Persistent illnesses are those diseases or disabilities that were experienced for more than 30 days in the preceding year. Persistent illnesses include chronic fever, disability or injury, heart disease, respiratory disease or asthma or bronchitis, gastric or ulcer, hypertension, diabetes, cancer, paralysis, kidney-related diseases, liver-related diseases, arthritis or rheumatism, skin problems, and eye problems to name a few.

Limitations of the study

The survey data were collected on the incidence of selected diseases during monsoon and dry seasons from a representative sample of households. Although the survey is representative of both urban and rural areas, the linearized error terms of some of the illnesses exceed the recommended level of dispersion— more than 15 percent. It is not representative at the suboutcome level. A larger sample size could have helped for robustness of the findings as well as to enhance their reliability.

The study presents some other limitations that should be investigated in future research to estimate the burden of disease. Not all illnesses were medically diagnosed, and disease categories were inferred from a wide array of symptoms reported by the respondents. Also, the severity of disease or symptoms was not considered. Prevalence of air pollution, which is directly linked to several diseases, including respiratory diseases, cardiovascular damage, fatigue, headaches, and anxiety, has not been considered. Finally, correlations with different health outcomes is not evaluated. This is particularly important for malnutrition, which is related to the occurrence of diseases.

TABLE 3.3 **Categorization of infectious diseases**

CLASSIFICATION	DIAGNOSED DISEASES	SYMPTOMS CLASSIFIED WITH DISEASES BUT UNDIAGNOSED
Common cold	Common cold	Fever with runny nose, chills, sore throat
		Runny nose, chills, sore throat, cough (without fever)
Vector-borne diseases	Dengue (classical)	Fever with body aches, pain in small joints, retro-orbital pain, rash
	Dengue (hemorrhagic)	Fever with body aches, chills, pain in small joints, retro-orbital pain, rash, hemorrhage
	Chikungunya	Fever with body aches, joint pain, radiating joint pain
	Malaria	Fever with chills, body aches
Respiratory illnesses	Pneumonia	
	Influenza-like illness	Fever with cough, sore throat, body aches, headache
	Severe acute respiratory infection	Fever with cough, sore throat, body aches, headache, breathing difficulty
Waterborne diseases	Diarrhea	
	Dysentery	

Source: Original table for this publication.
Note: Table shows the classification of infectious diseases (first column). The middle column shows a list of medically diagnosed illnesses reported by the respondents. The last column on the right shows symptoms that were grouped together—with the Institute of Epidemiology and Disease Control Research's guidance—and assigned to the disease category for patients who did not seek medical care.

Climate variables used in this study are based on data collected from only 43 weather stations of the BMD. For enumeration areas where a weather station is not available, the BMD provided the closest approximation. This may have resulted in imprecisions in the measurement of local climate conditions and, consequently, in the estimated impact of climate variability on the incidence of diseases.

The scope of the questionnaire was broadened to meet the needs of users. The structured questionnaire contained 15 pages along with some open-ended questions. As a result, some of the respondents were unwilling to give adequate time to the interviewers, and this may have affected their responses and consequently some of the findings. For questions regarding expenditures, respondents were not asked to provide any documentary evidence, and some had difficulty recalling the amounts of money spent.

NOTE

1. Questions ask whether or the degree to which the respondent or others in the household face difficulty in performing basic universal activities such as moving, seeing, hearing, cognition, self-care, and communication (Washington Group 2018).

REFERENCES

Arafat, S. M. Y., A. Shahoriar, A. Faruq, Z. Sojib Bin, and A. Amin. 2018. "Depression in Spinal Cord Injury Patients: A Cross-sectional Observation with PHQ-9 in a Rehabilitation Center of Bangladesh." *Journal of Behavioral Health* 7 (1): 36.

Bangladesh, NIPORT (National Institute of Population Research and Training), Mitra and Associates, and ICF. 2016. *Bangladesh Demographic and Health Survey 2014*. Dhaka: NIPORT. https://dhsprogram.com/pubs/pdf/FR311/FR311.pdf.

Chowdhury, A. N., S. Ghosh, and D. Sanyal. 2004. "Bengali Adaptation of Brief Patient Health Questionnaire for Screening Depression at Primary Care." *Journal of the Indian Medical Association* 102 (10): 544–47.

Fouque, F., and J. C. Reeder. 2019. "Impact of Past and Ongoing Changes on Climate and Weather on Vector-Borne Diseases Transmission: A Look at the Evidence." *Infectious Diseases of Poverty* 8: 51.

Hossain, S., A. Anjum, M. E. Uddin, M. A. Rahman, and M. F. Hossain. 2019. "Impacts of Socio-Cultural Environment and Lifestyle Factors on the Psychological Health of University Students in Bangladesh: A Longitudinal Study." *Journal of Affective Disorders* 256: 393–403. doi:10.1016/j.jad.2019.06.001.

Islam, S., R. Akter, T. Sikder, and M. D. Griffiths. 2020. "Prevalence and Factors Associated with Depression and Anxiety among First-Year University Students in Bangladesh: A Cross-Sectional Study." *International Journal of Mental Health and Addiction.* doi:10.1007/s11469-020-00242-y.

Islam, S. M. S., L. B. Rawal, and L. W. Niessen. 2015. "Prevalence of Depression and Its Associated Factors in Patients with Type 2 Diabetes: A Cross-sectional Study in Dhaka, Bangladesh." *Asian Journal of Psychiatry* 17: 36–41. doi:10.1016/j.ajp.2015.07.008.

Levis, B., A. Benedetti, and B. D. Thombs. 2019. "Accuracy of Patient Health Questionnaire-9 (PHQ-9) for Screening to Detect Major Depression: Individual Participant Data Meta-Analysis." *BMJ* 365: l1476. doi:10.1136/bmj.l1476.

Mamun, M. A., N. Huq, Z. F. Papia, S. Tasfina, and D. Gozal. 2019. "Prevalence of Depression among Bangladeshi Village Women Subsequent to a Natural Disaster: A Pilot Study." *Psychiatry Research* 276: 124–28. doi:10.1016/j.psychres.2019.05.007.

Manea, L., S. Gilbody, and D. McMillan. 2015. "A Diagnostic Meta-analysis of the Patient Health Questionnaire-9 (PHQ-9) Algorithm Scoring Method as a Screen for Depression." *General Hospital Psychiatry* 37 (1): 67–75. doi:10.1016/j.genhosppsych.2014.09.009.

Mani, M., and L. Wang. 2014. "Climate Change and Health Impacts: How Vulnerable Is Bangladesh and What Needs to Be Done." *End Poverty in South Asia* (blog), May 7, 2014. https://blogs.worldbank.org/endpovertyinsouthasia /when-climate-becomes-health-issue-how-vulnerable-bangladesh.

Moonajilin, M. S., M. E. Rahman, and M. S. Islam. 2020. "Relationship between Overweight/ Obesity and Mental Health Disorders among Bangladeshi Adolescents: A Cross-sectional Survey. *Obesity Medicine* 18: 100216. doi:10.1016/j.obmed.2020.100216.

Roy, T., C. E. Lloyd, M. Parvin, K. G. B. Mohiuddin, and M. Rahman. 2012. "Prevalence of Co-morbid Depression in Out-patients with Type 2 Diabetes Mellitus in Bangladesh." *BMC Psychiatry* 12 (1): 123. doi:10.1186/1471-244X-12-123.

Spitzer, R. L., K. Kroenke, and J. B. Williams. 1999. "Validation and Utility of a Self-Report Version of PRIME-MD: The PHQ Primary Care Study. Primary Care Evaluation of Mental Disorders, Patient Health Questionnaire." *JAMA* 282 (18): 1737–44. doi:10.1001/jama .282.18.1737.

Spitzer, R. L., K. Kroenke, J. B. W. Williams, and B. Löwe. 2006. "A Brief Measure for Assessing Generalized Anxiety Disorder: The GAD-7." *Archives of Internal Medicine* 166 (10): 1092–97. doi:10.1001/archinte.166.10.1092.

Washington Group. 2018. "Washington Group Short Set of Questions on Disability." Washington, DC: Centers for Disease Control and Prevention. https://www.cdc.gov/nchs /washington_group/wg_questions.htm.

4 Results and Analyses

PATTERNS OF CLIMATE VARIABILITY PRIOR TO THE SURVEY

This first section of chapter 4 presents climatic conditions related to the survey covering the monsoon season from August to September 2019 and the dry season from January to February 2020. Three key weather variables—maximum temperature, minimum temperature, and rainfall—feature in the data from two months preceding each round of survey; they are presented in figure 4.1. These weather variables are reported for Dhaka, Chattogram, and national averages for the months of May, June, November, and December 2019. Between May and June, the maximum temperature decreased considerably across the country, while the minimum temperature increased slightly. Rainfall reported for Dhaka was substantially higher than in Chattogram, which is unusual because historical analysis of weather data shows rainfall in Chattogram is usually higher than in Dhaka. While rainfall increased between May and June for Dhaka and nationally, it decreased for Chattogram. The weather variables for November and December 2019 show usual patterns of temperature—both maximum and minimum—falling significantly compared to temperatures in May and June, while no rainfall was recorded in these two months.

The heat indexes for Dhaka and Chattogram are presented in figure 4.2. Heat index is a measure of "real feel" that combines relative humidity and actual air temperature (United States, NWS 2020). Evidence from Bangladesh indicates a higher risk of cholera two days after heat waves during the rainy season (Wu et al. 2018) and that heat waves can increase the risk of transmission of Nipah virus to humans as the bats that carry the Nipah virus are under physiological stress with extreme heat conditions, which could trigger prolonged viral shedding (Rahman et al. 2019).

Given the potential impact of heat waves on health conditions—not just infectious diseases—a heat index was constructed for the months of May, June, July, August, November, and December 2019 and January and February 2020. May to August 2019 covers the period of data collection during the monsoon as well as two months prior to it, while November 2019 to February 2020 covers the dry season data collection as well as two months prior. No variation presents in

FIGURE 4.1
Average weather variables, two months preceding each of the two rounds of surveys, 2019–20

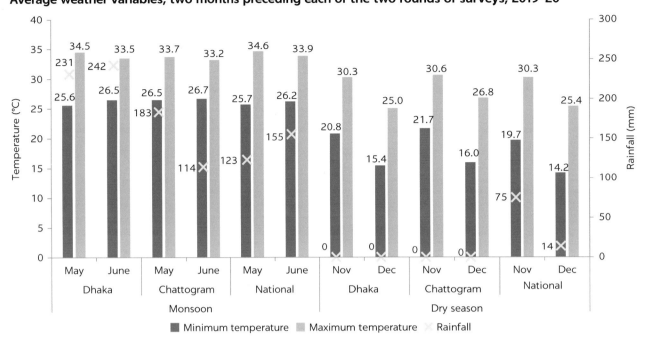

Source: Original figure for this publication.

FIGURE 4.2
Heat index measured in degrees Celsius, 2019–20

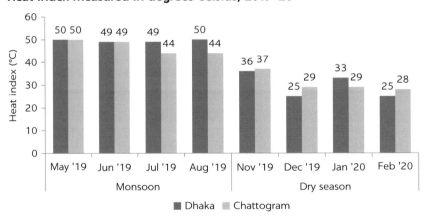

Source: Original figure for this publication.
Note: Heat index is a measure of "real feel" that combines relative humidity and actual air temperature (United States, NWS 2020).

the heat indexes of May and June 2019 for Dhaka and Chattogram cities. However, in July and August 2019, it is higher for Dhaka city than Chattogram city by 5°C to 6°C. The heat index for Dhaka city during the dry season was lower than Chattogram city, except for January 2020.

INFECTIOUS DISEASES

The subsections below analyze illness patterns and health-seeking behaviors across the two seasons covered by the survey—the monsoon when data was collected in July and August 2019 and the dry or post-monsoon of January and February 2020. The results are presented with varying degrees of disaggregation across areas of geographic representation, demographic and socioeconomic status, and household water, sanitation, and hygiene (WASH) practices. As Lowe et al. (2017) indicate, both rainfall and drought can increase the availability of potential habitats for mosquitoes—containers with stagnant clean water—and, therefore, availability of adequate WASH facilities is an important compounding factor for dengue, among others.

Prevalence of infectious diseases

This subsection discusses the prevalence of infectious diseases defined for individuals who have reported experiencing any infectious diseases or illness within 30 days preceding the survey across the monsoon and dry seasons. These infectious diseases, excluding the common cold, are classified as vector-borne, waterborne, or respiratory diseases for the purposes of analyses. Disease-specific figures are presented as a proportion of reporting any infectious disease, excluding the common cold, within these three categories. The common cold is separately reported.

Overall, the prevalence of the common cold is more than other infectious diseases across seasons and location, except for Dhaka and Chattogram cities in the monsoon. On average, the likelihood of reporting any infectious disease, excluding common cold, is 1.2 percentage points higher at the national level during monsoon than in the dry season, 5.7 versus 4.5 percent (figure 4.3). While overall urban and rural areas are generally comparable and remain so across the seasons, Dhaka and Chattogram cities report the highest number of infectious diseases, excluding the common cold, across the seasons.

FIGURE 4.3

Prevalence of any illness, by season, 2019–20

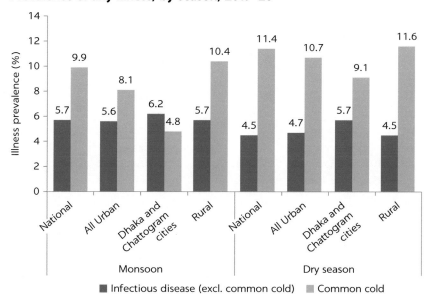

Source: Original figure for this publication.
Note: The figure shows weighted averages across geographical clusters.

Respiratory illnesses are the highest reported—62 percent of all morbidity—across seasons nationally. At the subnational level, while rates are comparable across all urban and rural areas and over seasons at approximately 61 percent, rates in Dhaka and Chattogram cities are the lowest during monsoon at 43.8 percent and highest during the dry season, at 66 percent. Figure 4.4 provides the breakdown of any illness by categories across the seasons.

Vector-borne diseases are more prevalent in Dhaka and Chattogram cities across the two seasons, monsoon and dry, compared to the national, all urban, and rural areas (figure 4.4). Subnationally, vector-borne diseases are reported by 22 percent of all urban areas and 25 percent in rural areas during the monsoon. The rates are the highest in Dhaka and Chattogram cities at 34 percent. During the dry season, while the rates are consistent across all urban and rural areas—between 14 and 15 percent—prevalence of vector-borne diseases remain the highest in Dhaka and Chattogram cities at almost 20 percent. Prevalence of vector-borne diseases is lower in the dry season compared to the monsoon across all locations.

Finally, the lowest proportion of those ill report contracting waterborne diseases (figure 4.4). At the national level, while only 14 percent are infected with a waterborne disease during the monsoon, the proportion increases to 23 percent during the dry season. Subnationally, while rates across all urban and rural areas remain comparable to each other, the rates increase by approximately 10 percentage points between the two seasons—15 and 14 percent, respectively, during monsoon versus 23 percent during the dry season. Dhaka and Chattogram cities deviate from the rest of the country across the seasons with respect to

FIGURE 4.4

Prevalence of vector-borne, waterborne, and respiratory diseases in monsoon and the dry season, 2019–20

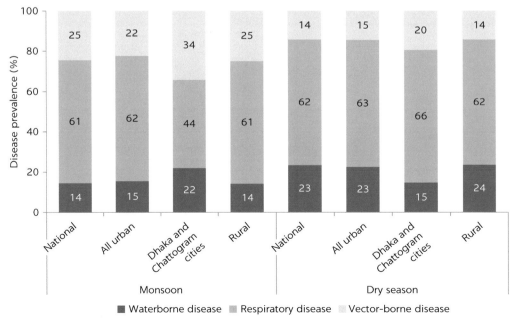

Source: Original figure for this publication.
Note: The categories of infectious disease (vector-borne, waterborne, and respiratory diseases) are a subset of contracting any infectious disease (excluding the common cold) in the 30 days before a survey.

waterborne diseases. While the prevalence is higher during monsoon in these cities at 22 percent compared to national, all urban, and rural areas, it reduces significantly in the dry season to 15 percent and is the lowest among all the geographical areas.

The data on disease prevalence is alternatively presented as a proportion of the total population as opposed to a subgroup of those who reported an illness to demonstrate the scale of morbidity. As shown in table 4.1, vector-borne and respiratory diseases are significantly higher in the Dhaka and Chattogram cities than in rural areas in the dry season.

Figure 4.5 shows prevalence of infectious diseases and common cold separately across age groups. Nearly one in ten children under 5 years of age and the elderly age 65 or more years report a seasonal illness, excluding the common

TABLE 4.1 **Infectious diseases, excluding the common cold, as a proportion of the total sample, 2019–20**

percent

	MONSOON					DRY SEASON				
DISEASE CATEGORY	ALL URBAN	DHAKA AND CHATTOGRAM CITIES	RURAL	DIFFERENCE		ALL URBAN	DHAKA AND CHATTOGRAM CITIES	RURAL	DIFFERENCE	
	(1)	(2)	(3)	(3) – (1)	(3) – (2)	(4)	(5)	(6)	(6) – (4)	(6) – (5)
Waterborne	0.9	1.4	0.8	n.a.	*	1.1	0.8	1.1	n.a.	n.a.
Respiratory	3.5	2.7	3.5	n.a.	n.a.	2.9	3.8	2.8	n.a.	*
Vector-borne	1.3	2.1	1.4	n.a.	n.a.	0.7	1.1	0.6	n.a.	*
N	15,409					14,758				

Source: Original table for this publication.

Note: Table shows geographic prevalence of diseases across seasons. Figures presented are weighted means. Test (columns labeled as "Difference") show significance levels from a weighted *t*-test. n.a. = not applicable.

* $p < 0.10$.

FIGURE 4.5

Prevalence of infectious diseases across age groups, 2019–20

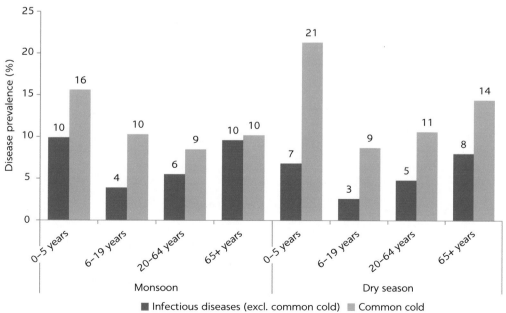

Source: Original figure for this publication.

Note: Figure shows weighted means across age groups and seasons.

cold, during the monsoon. The likelihood of doing so is the lowest among adolescents ages 6 through 19 years, at 4 percent, followed by adults of between 20 and 64 years, at 6 percent in the same season. Though overall rates of morbidity go down during the dry season, they remain proportional to the rates in the monsoon. In the dry season, the elderly of ages 65 or more years report the highest incidence of infectious diseases, excluding the common cold.

Figure 4.6 presents the prevalence of infectious diseases by category. Respiratory diseases remain most pervasive across the two seasons, with prevalence significantly higher during the dry season, except for children—ages 0 to 5 years—in the dry season when they experience more waterborne diseases.

Prevalence of respiratory illnesses is the highest among the elderly at 72 percent in the monsoon and 83 percent in the dry season. Nearly 65 percent of children of ages 5 years or under report a respiratory disease during the monsoon season while the number markedly reduces to 37 percent during the dry period. The prevalence of respiratory illnesses among the younger population of ages 6 to 19 years and adults of ages 20 to 64 years remains comparable at 59 percent during monsoon, although it increases considerably for both groups during the dry season.

In the monsoon, reported rates of waterborne diseases are highest among children of ages 5 years or less and the elderly of ages 65 years or more—22 and 24 percent, respectively—while the lowest for younger populations from ages 6 to 19 years at 6 percent. Overall, rates of waterborne diseases increase for all during the dry season, with the exception of people ages 65 years and above. Incidences increase by nearly threefold to 56 percent among children ages

FIGURE 4.6

Prevalence of infectious diseases, excluding the common cold, by category, across age groups, 2019–20

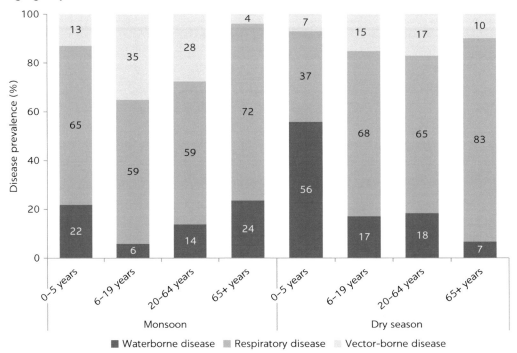

Source: Original figure for this publication.
Note: Figure shows weighted averages across age groups. The categories of infectious disease excluding common cold (vector-borne, waterborne, and respiratory diseases) are a subset of contracting any infectious disease in the 30 days preceding the survey.

5 years or less in the dry season. Prevalence among the younger population from ages 6 to 19 years, and the adults from ages 20 to 64 years remain comparable, between 17 and 18 percent in the dry season (figure 4.6).

The prevalence of vector-borne diseases is the highest among the younger population from, ages 6 to 19 years, at 35 percent, followed by adults from ages 20 to 64 years at 28 percent during the monsoon (figure 4.6). The rate is 13 percent among children of ages 5 years or less and the lowest among the elderly of ages 65 years or more (4 percent) during monsoon. Prevalence is generally lower during the dry season, with the exception of the elderly (ages 65 years or more). As in the monsoon, the younger population (from ages 6 to 19 years) and the adults (from ages 20 to 64 years) reported the highest rates (15 and 17 percent, respectively).

Figure 4.7 presents the likelihood of infectious diseases across socioeconomic quintiles. Morbidity declines with increasing socioeconomic status and the

FIGURE 4.7

Distribution of infectious diseases across socioeconomic status, any illness, and by illness category, 2019–20

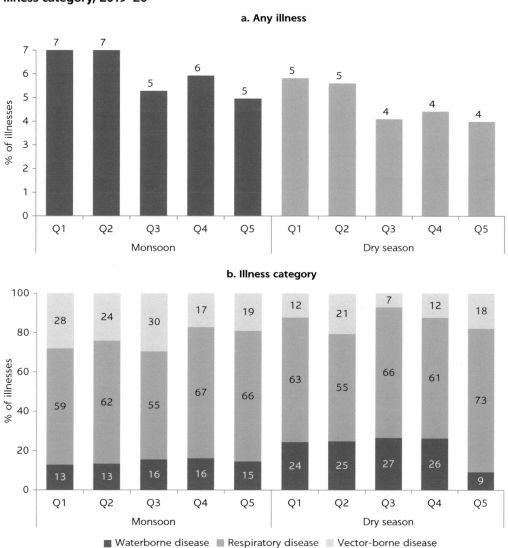

Source: Original figure for this publication.

Note: The figure shows weighted averages across socioeconomic quintiles constructed using an asset and facilities index. The categories of illnesses (vector-borne, waterborne, and respiratory diseases) exclude common cold and are a subset of contracting any illness in the last 30 days. Q1 represents the poorest quintile and Q5, the richest.

trend holds across the monsoon and the dry seasons. Trends in disease types across socioeconomic status reflect intuitive trends. For instance, respiratory diseases are the most reported, particularly during the dry season. The highest rates of respiratory-related morbidity are reported by the highest quintile. Compared to the monsoon, the difference between the bottom and top quintiles increase to 10 percentage points in the dry season.

On average, the likelihood of reporting a waterborne disease is lower during monsoon than in the dry season. The difference in the prevalence rates is limited between quintiles during monsoon, ranging between 13 and 16 percent. The overall prevalence rates increase by up to 10 percentage points during the dry season, except for the richest quintile where it decreases. The rates of waterborne diseases among the richest are significantly lower in comparison to the poorest, 9 versus 24 percent.

Vector-borne diseases are more prevalent among lower socioeconomic groups in monsoon—28 percent among the poorest in contrast to 19 percent among the richest. While the prevalence rates are marginally lower during the dry season, the most dramatic drops are experienced by those in the lower three quintiles—17 percentage points on average. Proportions among the richest remain comparable across the seasons; this could be because data collection during monsoon overlapped with Eid holidays when a significant proportion of the rich travel abroad and thus may have been less exposed to vector-borne diseases.

Figure 4.8 shows the concentration of illnesses across socioeconomic status during monsoon and the dry season. Concentration curves are used to identify

FIGURE 4.8

Equality of illnesses across socioeconomic status, monsoon, and dry season, 2019–20

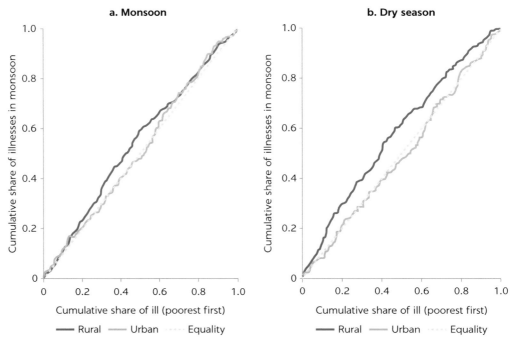

Source: Original figure for this publication.
Note: The figure shows concentration of illnesses across the range of a household's relative socioeconomic status, stratified by urban and rural areas.

whether socioeconomic inequality in the share of illnesses exists and whether it is more pronounced at one time than another. The curves represent the cumulative share of illnesses according to the cumulative share of population, ranked from poorest to the richest. The further a curve is above the reference equality 45-degree line, the more the corresponding health variable, that is share of illnesses, is concentrated among the poorest households. When the concentration curve lies under the 45-degree line, the corresponding health variable is more concentrated among the richest households.

While no differences exist across the socioeconomic percentiles during monsoon, the concentration curve corresponding to the rural areas lies above the 45-degree line during the dry season, suggesting that the rates of morbidity are more concentrated among the poor during this time. The concentration curve for the urban areas during the dry season is very close to the 45-degree line, which indicates very little association between share of illnesses and socioeconomic status.

Correlates of infectious diseases

Table 4.2 shows correlates of contracting any infectious disease, excluding common cold. Coefficients presented are derived using a weighted linear specification. The likelihood of contracting an infectious disease is 19.7 percentage points lower in the dry season than during monsoon. Overall, younger people of ages 6 to 19 years and adults of ages 20 to 64 years are 5.0 and 3.9 percentage points less likely to experience an infectious disease than children of ages 5 years or less or the elderly 65 years or more. Seasonal differentials suggest younger people from

TABLE 4.2 **Correlates of contracting any seasonal illness, excluding the common cold, 2019–20**

	COEFFICIENT	STANDARD ERROR
Season (1 = dry, monsoon = 0)	−0.197***	0.034
Age (base: 0 to 5 years)		
6 to 19 years	−0.050***	0.007
20 to 64 years	−0.039***	0.006
65+ years	−0.011	0.010
Age (base: 0 to 5 years) *(in dry season compared with monsoon)*		
6 to 19 years (dry season)	0.016*	0.009
20 to 64 years (dry season)	0.020**	0.008
65+ years (dry season)	0.010	0.014
Gender (1 = male, 0 = female)	0.005*	0.003
Individual is disabled (1 = y, 0 = n)	0.018***	0.004
Never married (1 = y, 0 = n)	−0.008*	0.004
Education (continued years)	−0.001**	0.000
Male head of household (1 = y, 0 = n)	−0.011**	0.005
Household size	−0.005***	0.001

continued

TABLE 4.2, *Continued*

	COEFFICIENT	STANDARD ERROR
Socioeconomic quintiles (base: bottom two quintiles [Q])		
Q3	−0.015***	0.004
Q4	−0.008**	0.004
Q5 (richest)	−0.010**	0.005
WASH index (base: lowest)		
WASH index (mid)	−0.012***	0.003
WASH index (high)	−0.009**	0.004
Household experienced a shock (1 = y, 0 = n)	0.010***	0.003
Community characteristics		
Number of households in community_ihs	0.064	0.106
Trash deposited in a single location (1 = y, 0 = n)	−0.063*	0.034
Connectivity index (base: least)		
Connectivity index (mid)	0.005	0.059
Connectivity index (high)	−0.085*	0.046
Average hours of electricity available_ihs	−0.438	0.649
Location (base: rural areas)		
Cities of Dhaka and Chattogram	−0.053	0.067
Cities of Dhaka and Chattogram (in dry season compared with monsoon)	0.018	0.012
Weather (means from two months preceding survey)		
Humidity (two-month mean, presurvey)	−0.002*	0.001
Temperature (two-month mean, presurvey)	−0.014***	0.003
$R^2 = 0.03$		
$N = 30,091$		

Source: Original table for this publication.
Note: The table shows coefficients from a weighted linear specification. The dependent variable is whether the individual experienced any seasonal illness. Variables with suffix "_ihs" refer to inverse hyperbolic sine transformations. The models account for primary-sampling-unit-level heterogeneity.
*$p < 0.10$; **$p < 0.05$; ***$p < 0.01$.

ages 6 to 19 years are 1.6 percentage points more likely to suffer from an infectious disease during the dry season, while the similar is true for adults of ages 20 to 64 years, at 2 percentage points. Overall, women are 0.5 percentage points less likely to suffer from an infectious disease than men. The likelihood of an infectious disease is 1.8 percentage points higher among disabled individuals than not. Similarly, individuals who never married are 0.8 percentage points less likely to suffer infectious diseases.

An additional year of education is negatively correlated with the probability of having an infectious disease at 0.1 percentage points. Regarding household characteristics, residing in a household with a male head is negatively correlated with illness at 1.1 percentage points. Similarly, a larger household size is negatively correlated with morbidity at 0.5 percentage points. In terms of socioeconomic status, a higher wealth status is negatively correlated with morbidity. The same is true for a household's WASH index—placement in a higher quintile reduces the likelihood of contracting infectious diseases. Members of a household that experienced a shock in the 12 months preceding the survey are more likely to have an infectious disease at 1 percentage point. Households in neighborhoods that are more organized in the way they deal with their household refuse are less likely to experience an illness, at 6.3 percentage points. A household placed in a neighborhood with a higher connectivity index is negatively correlated with illnesses. Finally, weather is correlated with illnesses. An increase in humidity and mean temperature by one unit in two months preceding the survey reduces the likelihood of contracting an illness by 0.2 and 1.4 percent, respectively.

The condition-specific correlates presented in table 4.3 are also estimated using a linear specification. More specifically, table 4.3 explores factors that affect the likelihood of contracting waterborne diseases (column 1), respiratory infections (column 3), and vector-borne diseases (column 5) generated using a linear specification. While adolescents of ages 6 to 19 years and adults of ages 20 to 64 years are 16.9 and 8.3 percentage points, respectively, less likely to

TABLE 4.3 **Correlates of contracting vector-borne, waterborne, or respiratory infections, 2019–20**

	WATERBORNE DISEASES		RESPIRATORY INFECTIONS		VECTOR-BORNE DISEASES	
	COEFFICIENT	STD. ERROR	COEFFICIENT	STD. ERROR	COEFFICIENT	STD. ERROR
	(1)	(2)	(3)	(4)	(5)	(6)
Season (1 = dry season, monsoon = 0)	−0.124	0.247	0.337	0.313	−0.214	0.255
Age (base: 0 to 5 years)						
6 to 19 years (adolescents)	−0.169***	0.046	0.021	0.058	0.148***	0.047
20 to 64 years (adults)	−0.084**	0.038	−0.009	0.049	0.093**	0.040
65+ years (elderly)	0.035	0.056	0.101	0.071	−0.136**	0.058
Age (base: 0 to 5 years) (in dry season compared with monsoon)						
6 to 19 years (dry season)	−0.190***	0.064	0.278***	0.081	−0.087	0.066
20 to 64 years (dry season)	−0.280***	0.053	0.324***	0.067	−0.044	0.054
65+ years (dry season)	−0.495***	0.079	0.368***	0.100	0.127	0.081
Gender (1 = male, 0 = female)	−0.005	0.019	0.013	0.025	−0.008	0.020
Individual is disabled (1 = y, 0 = n)	0.038	0.026	−0.088***	0.033	0.050*	0.027
Never married (1 = y, 0 = n)	0.022	0.037	−0.056	0.046	0.034	0.038
Education (in years)	0.001	0.003	0.002	0.004	−0.003	0.003
Male head of household (1 = y, 0 = n)	−0.041	0.037	0.001	0.047	0.040	0.038
Household size	0.003	0.006	−0.002	0.008	−0.001	0.007
Socioeconomic quintiles (base: bottom two quintiles [Q])						
Q3	0.028	0.028	−0.048	0.036	0.020	0.029
Q4	0.049	0.031	−0.007	0.040	−0.042	0.032
Q5 (richest)	−0.049	0.036	0.083*	0.046	−0.034	0.037
WASH index (base: lowest)						
WASH index (mid)	0.006	0.027	−0.034	0.035	0.028	0.028
WASH index (high)	−0.035	0.033	0.023	0.042	0.012	0.034
Household experienced a shock (1 = y, 0 = n)	−0.005	0.024	−0.057*	0.031	0.062**	0.025

continued

	WATERBORNE DISEASES		RESPIRATORY INFECTIONS		VECTOR-BORNE DISEASES	
	COEFFICIENT	STD. ERROR	COEFFICIENT	STD. ERROR	COEFFICIENT	STD. ERROR
	(1)	(2)	(3)	(4)	(5)	(6)
Community characteristics						
Number of households in community_ihs	−0.682	0.608	0.625	0.772	0.057	0.628
Trash deposited in a single location (1 = y, 0 = n)	−0.279	0.297	−0.048	0.377	0.327	0.307
Connectivity index (base: least)						
Connectivity index (mid)	−0.240	0.341	0.417	0.434	−0.176	0.353
Connectivity index (high)	−0.186	0.348	0.600	0.442	−0.414	0.360
Average hours of electricity available_ihs	4.862	3.719	−6.092	4.726	1.230	3.842
Location (base: rural areas)						
Cities of Dhaka and Chattogram	0.306	0.408	−0.369	0.518	0.063	0.421
Cities of Dhaka and Chattogram (in dry season compared with monsoon)	−0.156*	0.086	0.194*	0.109	−0.038	0.089
Weather (means from two months preceding survey)						
Humidity (two-month mean, presurvey)	−0.016**	0.007	0.015*	0.009	0.000	0.007
Temperature (two–month mean, presurvey)	−0.042*	0.022	0.057**	0.028	−0.014	0.023
R^2	0.24		0.22		0.24	
N	1,604		1,604		1,604	

Source: Original table for this publication.
Note: Table shows coefficients from a weighted linear specification. Dependent variable is whether the individual experienced waterborne, respiratory, or vector-borne illnesses. Variables with suffix "_ihs" refer to inverse hyperbolic sine transformations. The models account for primary-sampling-unit-level heterogeneity. WASH = water, sanitation, and hygiene.
*$p < 0.10$; **$p < 0.05$; ***$p < 0.01$.

experience waterborne illnesses than children below 5 years of age, the opposite is true for vector-borne diseases, with children and adolescents of ages 6 to 19 years and adults of ages 20 to 64 years experiencing more than children under 5 years of age by 14.8 and 9.4 percentage points, respectively. The elderly ages 65 years or more are 13.6 percentage points less likely to contract a vector-borne disease than children under 5 years. There is some heterogeneity across seasons and age groups. Adolescents, adults, and the elderly are less likely to experience waterborne diseases during the dry season by 19, 28, and 49.5 percentage points, respectively, than during monsoon. In contrast, all three groups are more likely to experience respiratory illnesses during the dry season than in monsoon, with 27.8, 32.4, and 36.8 percentage point increases for adolescents, adults, and the elderly, respectively. Disabled individuals are 8.8 percentage points less likely to experience respiratory infections while 5 percentage points more likely to contract a vector-borne disease during the dry season than in monsoon.

While morbidity rates among households based in the Dhaka and Chattogram cities do not vary on average compared to other urban or rural locations, the likelihood of contracting waterborne diseases diminishes during the dry season, while in line with the literature, the chance of contracting a respiratory illness increases by 19.4 percentage points. With regard to weather outcomes, humidity and mean temperature are negatively correlated to waterborne diseases by 1.6 and 4.2 percent, respectively. The opposite is true for respiratory illnesses; an increase in humidity and mean temperature increase the likelihood of suffering from a respiratory illness by 1.5 and 5.7 percent, respectively.

Health-seeking behavior and associated expenditure

Respondents who reported contracting any season or temporary illness—infectious or common cold—were asked about their behavior for seeking health care services across the two seasons. More of the respondents who had had any seasonal or temporary illness sought services from a health facility than those who did not. On average, nationally about 40 percent sought services from a health facility, almost half of the respondents in all urban areas and more than one-third in rural areas, across the two seasons. Figure 4.9 shows the pattern of health-care-seeking behavior for seasonal illnesses. Proportions seeking care from pharmacies or unqualified doctors were comparable across the locations

FIGURE 4.9

Health-care-seeking behavior for seasonal illnesses (infectious disease or common cold) across location and socioeconomic status, 2019–20

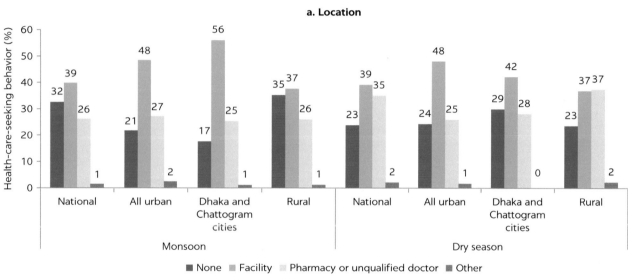

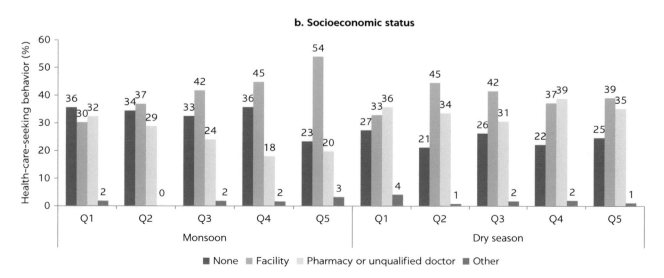

Source: Original figure for this publication.
Note: Figure shows weighted seasonal means and distribution of the type of health care sought for seasonal illnesses. Q1 is the poorest quintile and Q5, the richest.

and seasons, except for respondents in rural areas who sought care more from the pharmacies in the dry season than in the monsoon, 37 percent versus 26 percent, and nationally—35 percent in the dry season compared to 26 percent in monsoon.

Overall, more people accessed health care in the dry season compared to the monsoon across wealth quintiles. In the dry season, 27 percent of the poorest quintile did not access any health care compared to 36 percent in monsoon. In accessing health care services from a health facility, pharmacy, or unqualified doctor, no variation is evident between the two seasons for the poorest quintile, averaging about one in three people. For the richest quintile, accessing health care from a facility was the most popular option across the seasons, although the rate declined significantly between seasons—54 percent in monsoon compared to 39 percent in the dry season—as more people sought services from a pharmacy or unqualified doctor—20 percent in monsoon compared to 35 percent in the dry season. In the richest quintile, about a quarter of the respondents did not access any health care in both seasons—23 percent in monsoon and 25 percent in the dry season.

Figure 4.10 presents the type of facility accessed, as a proportion of respondents who reported accessing any facility (shown in figure 4.9). Overall, the use of private facilities or qualified doctors' private practice was the most common across locations, wealth status, and seasons. At the subnational level, use of government facilities was more in monsoon than in the dry season across all locations. Use of nongovernmental facilities was very limited across the locations and seasons.

Reliance on private facilities or qualified doctors' private practices tends to be higher with increasing socioeconomic status, and the trend holds during both the monsoon and dry seasons, except for the third quintile (Q3) in monsoon and Q4 in the dry season (figure 4.10). In the monsoon, 64 percent of the individuals in the richest quintile sought care from private facilities or qualified doctors' private practices compared to 54 percent by the poorest quintile. Similarly, in the dry season, 70 percent among the richest accessed private facilities compared to 47 percent among the poorest.

Between the monsoon and dry season, the proportion of the richest quintile accessing government facilities declined—38 percent in monsoon compared to 29 percent in the dry season. For the poorest quintile, the opposite holds, as more people from the poorest quintile report accessed a government facility in the dry season (51 percent) than in monsoon (45 percent). Using services at nongovernmental organization facilities tend to be the lowest across all socioeconomic groups in both the seasons, as shown in figure 4.10.

Figure 4.11 outlines the different dimensions of expenditure related to seeking health care for infectious diseases or the common cold across various locations. Overall, the majority of expenditures was purchasing medicines across both seasons and all locations—accounting for more than half of the total health-care-related expenditure. The total costs tend to be more across all locations in monsoon than the dry season as more people reported accessing a facility in monsoon than in the dry season (figure 4.9). Respondents in all urban areas, which include Dhaka and Chattogram cities, spent more in total than those living in rural areas across both seasons. An intriguing finding for Dhaka and Chattogram cities is that the total cost of health-care-related expenditure in the monsoon is lower than in the dry season, although the proportion of people accessing health care is higher in monsoon than in the dry season. A reason for

FIGURE 4.10

Type of facilities accessed for treament of infectious disease or common cold across location and socioeconomic status, 2019–20

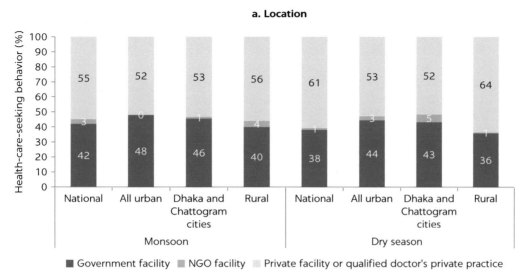

a. Location

Government facility ■ NGO facility ■ Private facility or qualified doctor's private practice

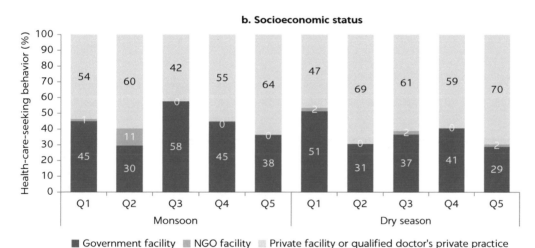

b. Socioeconomic status

Government facility ■ NGO facility ■ Private facility or qualified doctor's private practice

Source: Original figure for this publication.
Note: Figure shows distribution of the type of facility from which health care was sought for seasonal illnesses. Q1 represents the poorest quintile and Q5, the richest. NGO = nongovernmental organization.

this is that the proportion of people who access health care from a pharmacy or unqualified doctor in the dry season is higher than in monsoon—28 percent in the dry season compared to 25 percent in monsoon—which is reflected in significantly higher costs of consultation and medicines in the dry season.

Other direct costs such as pathological or laboratory tests accounted for the second-highest expenditure and were much higher than consultation fees across all locations and seasons. Indirect costs of health care such as transportation costs were the lowest as a proportion of total expenditure. Transportation costs were higher in monsoon that in the dry season for all locations, except the cities of Dhaka and Chattogram.

Health-care-related expenditure increases with socioeconomic status, and this trend holds in both the monsoon and dry seasons, except for the second

FIGURE 4.11

Health care expenses for seasonal illnesses, by location, 2019–20

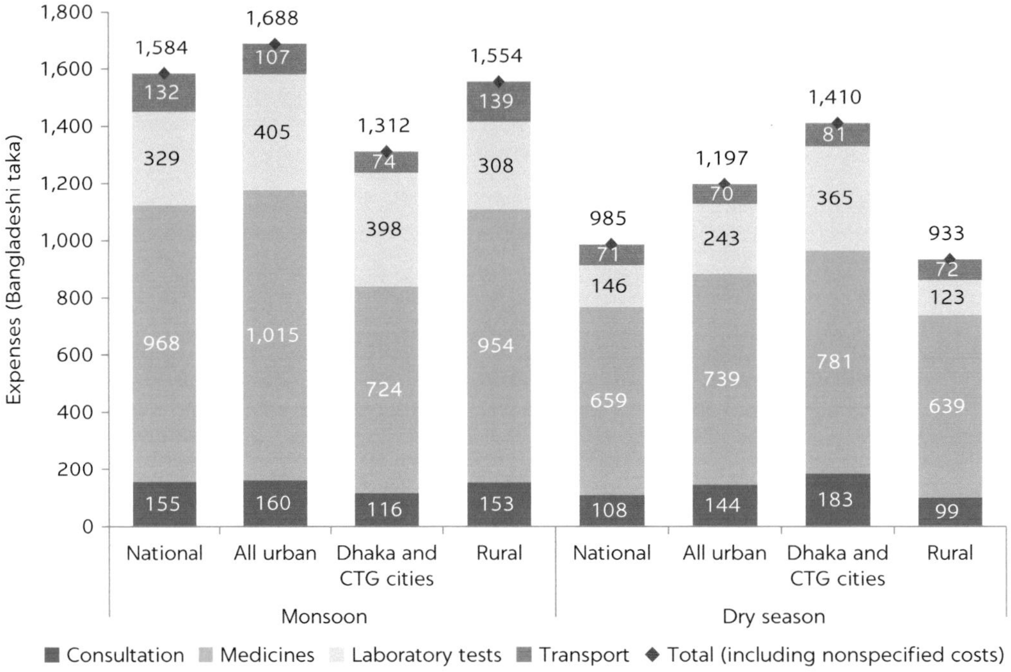

Source: Original figure for this publication.
Note: The figure shows weighted means of expenditures associated with seasonal illnesses in Bangladeshi taka, disaggregated by location across the seasons. CTG = Chattogram.

poorest quintile (Q2) in the monsoon (figure 4.12). Overall, total expenditure related to health care was lower in the dry season than monsoon across all wealth quintiles. Medicines and pathological or laboratory tests accounted for most of the health-care-related expenditures across all wealth quintiles in both seasons. Medicines accounted for more than half of the total cost for all wealth quintiles across seasons. In monsoon, the proportion of money spent on pathological or laboratory tests was lower than in the dry season.

PERSISTENT ILLNESSES

The following subsections analyze illness patterns and health-care-seeking behavior for chronic or persistent illnesses. As mentioned earlier, these are defined as diseases or disabilities that were experienced for more than 30 days in the 12 months preceding the survey. Since there is no variation in the prevalence of persistent or chronic illnesses by seasons, as expected, the results presented in this section are combined for the two rounds of survey.

Prevalence of persistent illnesses

On average, 14 percent of the population reports a persistent illness. The prevalence is highest in Dhaka and Chattogram cities at 18 percent, followed by all urban and rural areas at 16 and 14 percent, respectively, as shown in figure 4.13. In relation to demographics, adults of ages 20 to 64 years and the

FIGURE 4.12

Health care expenses for seasonal illnesses, by socioeconomic status, 2019–20

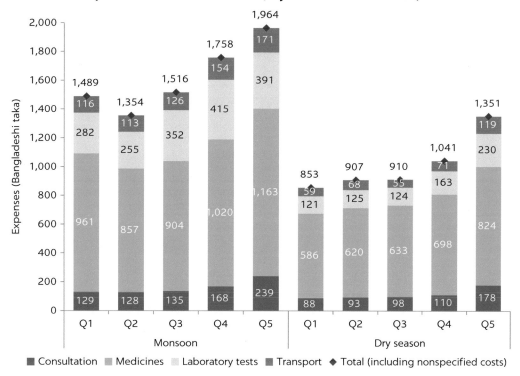

Source: Original figure for this publication.
Note: Figure shows weighted means of expenditures associated with seasonal illnesses, in Bangladeshi taka, disaggregated by asset quintiles. Q1 indicates the poorest quintile and Q5, the richest.

FIGURE 4.13

Prevalence of any persistent illness, by location, age group, and socioeconomic status, 2019–20

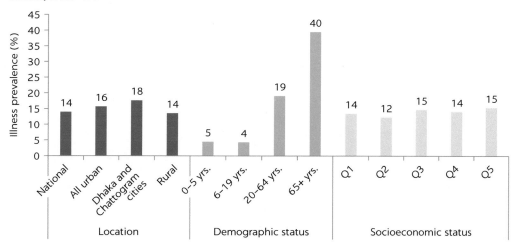

Source: Original figure for this publication.
Note: The figure shows weighted means of the prevalence of persistent conditions across geographic clusters, age groups, and socioeconomic status. Q1 indicates the poorest quintile and Q5, the richest quintile.

elderly ages 65 years or more report the most persistent illnesses—19 percent for adults and more than double that for the elderly. The prevalence does not vary significantly by wealth status.

Figure 4.14 presents disaggregated data for each category of illness by location, age group, and socioeconomic status. Noncommunicable diseases (NCDs) are reported more in all urban areas (76 percent) than in rural areas (71 percent). In rural areas, respiratory illnesses were more than in urban areas by 5 percentage points. Persistent fever accounted for a small proportion of chronic morbidities across all locations.

Prevalence of NCDs generally increases linearly by age. A higher proportion of adults from ages 20 to 64 years report having an NCD at 78 percent compared to the elderly at 72 percent. This finding is consistent with other nationally representative surveys that indicate an increasing proportion of the younger population of ages 35 years or more suffer from diabetes or hypertension. Children report the highest rates of persistent fever while the elderly report the lowest. On average, the prevalence of persistent respiratory illnesses is comparable across locations. Children ages 5 years or less and the elderly of ages 65 years or more are the most susceptible to persistent respiratory illnesses.

NCDs generally are most prevalent among individuals in the highest socioeconomic status, possibly driven by sedentary lifestyle. On the other hand, prevalence of persistent respiratory illnesses and fevers are negatively associated with the household's wealth status.

Health-seeking behavior and associated expenditure

Respondents who reported any persistent disease were asked about their behavior for seeking health care services (figure 4.15). The pattern of health-care-seeking behavior for persistent illnesses is different from the pattern observed for infectious diseases or common cold, as more people preferred not to seek

FIGURE 4.14

Prevalence of persistent illnesses, by location, age group, and socioeconomic status, 2019–20

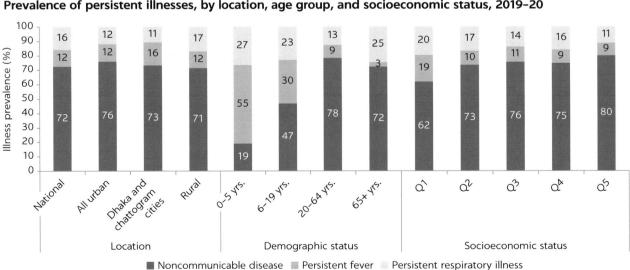

Source: Original figure for this publication.

Note: Figure shows weighted averages of the prevalence of persistent conditions across geographic clusters, age groups and socioeconomic status. The categories of illnesses—noncommunicable diseases, persistent fever, and persistent respiratory illnesses—are a subset of contracting any persistent illness that lasted for more than 30 days in the 12 months preceding the survey. Q1 represents the poorest quintile and Q5, the richest.

FIGURE 4.15

Health-care-seeking behavior for persistent illnesses, by location and socioeconomic status, 2019–20

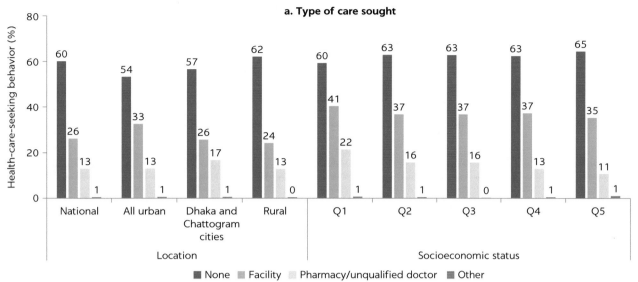

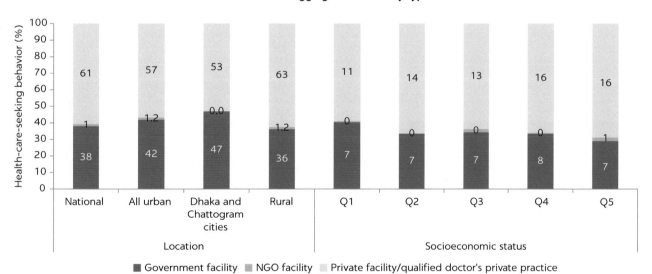

Source: Original figure for this publication.

Note: Figures show weighted means and distribution. Panel a shows whether the individual sought any care and whether it was informal or facility-based across location and socioeconomic status. Panel b shows disaggregation of facility-based care for the same illnesses by location and socioeconomic status. Q1 indicates the poorest quintile and Q5, the richest. NGO = nongovernmental organization.

health care across location and socioeconomic status. This is consistent with existing literature, which shows, for example, people with diabetes being not aware of their condition and hence not seeking care (Bangladesh NIPORT and ICF 2020).

The national average expenditure over a 12-month period of the individuals who reported a persistent illness is Tk 1,470 (Bangladeshi taka) (figure 4.16). Further disaggregation suggests that the costs are skewed toward urban areas—the largest figures are reported by those living in Dhaka and Chattogram cities averaging Tk 1,863, followed by those living in all urban areas at Tk 1,599. This suggests that the average in urban areas, excluding Dhaka and Chattogram cities,

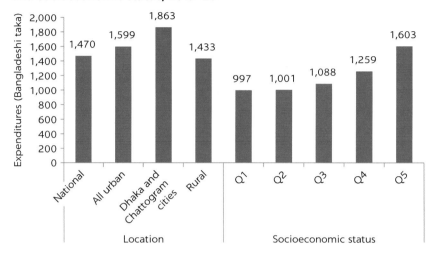

FIGURE 4.16

Health-care-related expenditure for persistent illnesses, by location and socioeconomic status, 2019–20

Source: Original figure for this publication.
Note: The figure shows weighted means of expenditures associated with persistent illnesses, in Bangladeshi taka, disaggregated by location and socioeconomic status. Q1 indicates the poorest quintile and Q5, the richest.

are more closely aligned with rural areas. Total cost of treatment also goes up with socioeconomic status.

In line with existing literature, a clear and positive relationship is found between health care expenses for persistent conditions and socioeconomic status (Raza et al. 2015). Those in the highest quintile (Q5), on average, spent 184 percent more for treating persistent conditions than those in the lowest quintile—Tk 1,603 versus Tk 997, respectively.

MENTAL HEALTH

Prevalence of depression and anxiety

Based on the same set of panel data used in the previous sections of this chapter, analyses of mental health issues are presented in this subsection. The analyses for depression are based on the Patient Health Questionnaire-9 (PHQ-9) score, which is a commonly used depression screening tool. Analyses for anxiety are based on the Generalized Anxiety Disorder-7 (GAD-7). A clinical cutoff score of 10 was used to establish the presence of depression and anxiety for both the PHQ-9 and the GAD-7, based on best-practice standards in the literature.

The weighted national prevalence of depression and anxiety was 16 percent and 6 percent, respectively. Prevalence of depression was the same across rural and urban locations at nearly 16 percent; the prevalence of anxiety, however, varied, with 7.9 percent in urban areas compared to 5.5 percent in rural areas. Prevalence estimates specific to age, gender, location, and seasons are presented in figure 4.17, which shows depression is more prevalent than anxiety across all these dimensions.

FIGURE 4.17

Prevalence of depression and anxiety, by location, demographics, and seasonality, 2019–20

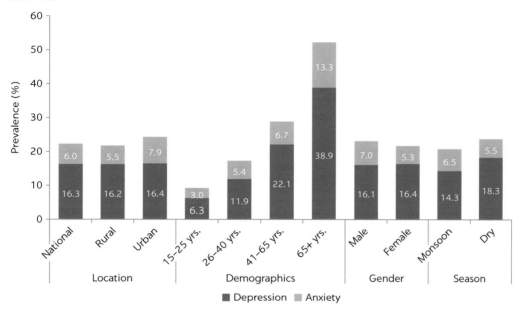

Source: Original figure for this publication.
Note: Depression analysis is based on Patient Health Questionnaire-9 scores and anxiety analysis on Generalized Anxiety Disorder-7.

Both depression and anxiety increase with age. Almost 40 percent of people age 65 years or more reported being depressed and 13 percent suffered anxiety. The prevalence of depression between male and female was the same, at approximately 16 percent, but men reported more anxiety than women. With respect to seasonality, more people were depressed in the dry season compared to monsoon—18 percent compared to 14 percent—while anxiety levels were similar with a difference of one percentage point.

Across all categories, the older age cohorts presented with more severe depressive symptoms (figure 4.18). Younger females of ages 26–40 years were more vulnerable to depression as were males 41 to 65 or more years in urban areas. In Dhaka and Chattogram cities, females of ages 26 to 40 years experienced more depression than other age groups. Representation for males and females in rural areas for depressive symptoms in the age group of 26 to 64 years was similar, while older males of ages 65 and more years experienced more depression compared to females in the same age cohort. Depressive symptoms were elevated during the dry season as compared to monsoon, with most pronounced increases among the two oldest cohorts. The pattern is slightly different for younger females, with more females of ages 26 to 40 years being depressed in the monsoon compared to the dry season.

Figure 4.19 shows the results for anxiety levels based on GAD-7 scores across the same parameters of age, gender, location, and age cohorts. The most pronounced difference was between urban and rural populations, with urban residents having substantially higher anxiety rates. Males in the age groups of 26 to 64 years appeared more anxious in rural areas, while in urban areas younger females in the age groups of 15 to 40 years were more anxious. During monsoon,

FIGURE 4.18

Prevalence of depression, by location, age group, gender, and season, 2019–20

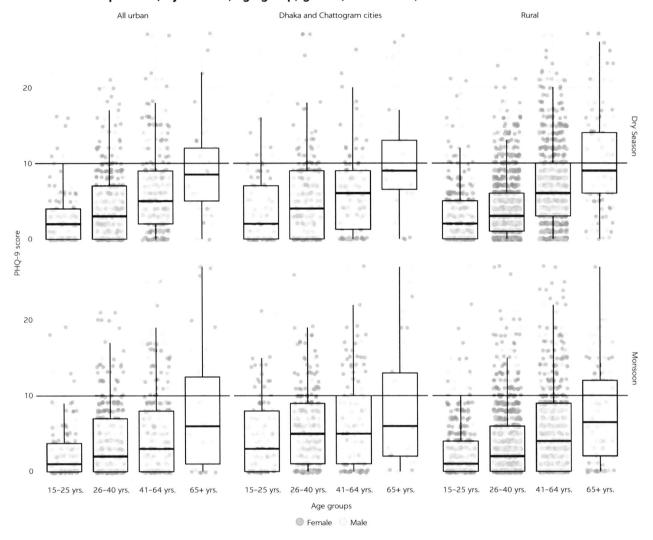

Source: Original figure for this publication.
Notes: The figure shows distribution of Patient Health Questionnaire-9 (PHQ-9) scores across locations (vertical axis) and seasons (horizontal axis). Results are further broken down by age groups (box plots). The horizontal lines within each box plot shows the median while the top and the bottom ranges show the 25th and 75th percentiles, respectively. The scatterplots represent gender (blue for male and pink for female). The line at score 10 represents the cutoff for the PHQ-9 beyond which the individual is likely to be associated with experiencing clinical depression.

females in Dhaka and Chattogram cities were more anxious than in the dry season. Anxiety levels for all urban and rural areas were similar in monsoon and the dry season.

The prevalence estimates of depression and anxiety in Bangladesh indicate that substantial morbidity of these mental disorders lie underaddressed, and the disorders are in urgent need of preventive and treatment initiatives. The most vulnerable for depression and anxiety are older, poorer, and disabled individuals. Accordingly, these populations should be accorded the highest priority for mental health services.

FIGURE 4.19

Prevalence of anxiety, by location, age group, gender, and season, 2019–20

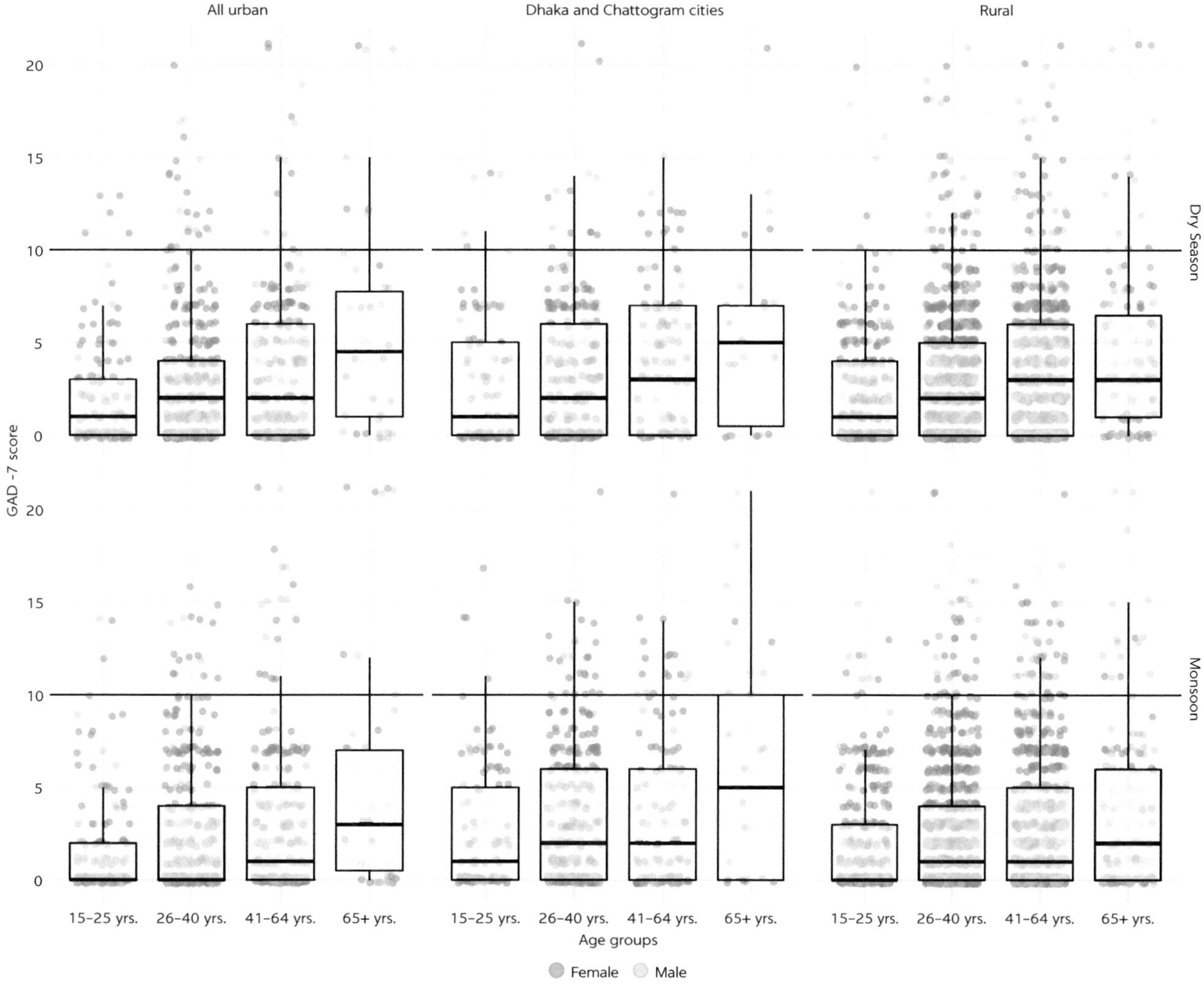

Source: Original figure for this publication.
Note: The figure shows distribution of Generalized Anxiety Disorder-7 (GAD-7) scores across locations (vertical axis) and seasons (horizontal axis). Results are further broken down by age groups (box plots). The horizontal lines in each box plot shows the median while the top and the bottom ranges show the 25th and 75th percentiles, respectively. The scatterplots represent gender (blue for male and pink for female). The line at score 10 represents the cutoff for the GAD-7 beyond which the individual is likely to be associated with experiencing clinical anxiety.

Correlates of depression and anxiety

Correlates of depression and anxiety are presented in table 4.4. Figures are marginal effects from an adjusted logit model. The models were informed by social determinants (Allen et al. 2014) and a biopsychosocial approach (Engel 1977) to mental illness in determining the inclusion of predictor variables. Accordingly, the study used the gender, age, and marital status of the respondent, as well as the gender of the household head, household size, education level, urban or rural location, and socioeconomic status as measured by asset quintiles. Due to sluggish progress and continued vulnerability of the Bangladeshi population to sudden economic, health, or environmental shocks, these measures, or shocks, as predictors of mental health have also been included. Any recent illness is also considered, as this outcome can impact the family members' mental health and

TABLE 4.4 **Correlates of depression and anxiety, 2019–20**

	DEPRESSION		ANXIETY	
	COEFFICIENT	STD. ERROR	COEFFICIENT	STD. ERROR
Season (1 = dry season, 0 = monsoon)	−0.177	0.109	0.073	0.066
Age (base: 15 to 25 years)				
26 to 40 years	0.052***	0.017	0.021**	0.010
41 to 64 years	0.096***	0.019	0.019*	0.011
65+ years	0.163***	0.024	0.041***	0.013
Gender (1 = male, 0 = female)	−0.020*	0.011	0.011*	0.006
Education (in years)	−0.006***	0.001	−0.001	0.001
Never married (1 = y, 0 = n)	0.013	0.027	−0.014	0.012
Location (1 = urban, 0 = rural)	0.004	0.012	0.022***	0.008
Male head of household (1 = y, 0 = n)	−0.013	0.020	0.003	0.008
Household size	0.001	0.003	−0.001	0.002
Individual is disabled (1 = y, 0 = n)	0.140***	0.018	0.034***	0.010
Any illness (1 = y, 0 = n)	0.057***	0.021	0.005	0.011
Socioeconomic quintiles (base: poorest quintile)				
Q2 (poorer)	−0.027*	0.016	−0.020**	0.008
Q3 (middle)	−0.039**	0.016	−0.021***	0.008
Q4 (richer)	−0.048***	0.016	−0.025***	0.009
Q5 (richest)	−0.061***	0.020	−0.030***	0.011
WASH index	−0.010***	0.003	−0.002	0.002
Household experienced a shock (1 = y, 0 = n)	0.046***	0.012	0.023***	0.007
Community characteristics				
Number of households in community	0.000***	0.000	0.000***	0.000
Trash deposited in a single location (1 = y, 0 = n)	0.033***	0.012	0.035***	0.007
Connectivity index (base: least)				
Connectivity index (mid)	0.019*	0.011	0.004	0.007
Connectivity index (high)	−0.020	0.016	−0.032***	0.007
Average hours of electricity available	0.000	0.001	0.000	0.001
Weather (means from two months preceding survey)				
Humidity (two-month mean presurvey)	0.000	0.002	0.003***	0.001
Temperature (two-month mean presurvey)	−0.016*	0.009	0.008*	0.005
Pseudo R^2	0.13		0.08	
N	7,086		7,086	

Source: Original table for this publication.
Note: The table shows marginal effects from weighted logit models. Dependent outcomes are whether an individual suffers from depression or anxiety. Models account for primary-sampling-unit-level heterogeneity. WASH = water, sanitation, and hygiene.
*$p < 0.10$; **$p < 0.05$; ***$p < 0.01$.

well-being. Disability is a strong predictor for developing mental health comorbidity, and accordingly was included in the models. In addition, seasonality—monsoon and dry season—has been incorporated to explore any effects on mental health and controlled for specific weather factors such as humidity and temperature. Finally, the models control for individual WASH practices and community characteristics.

Depression

Increasing age is significantly associated with increased probability of depression, with older populations substantially at higher risk (table 4.4). Compared to the youngest age group of 15 to 25 years, individuals between 26 and 40 years are more likely to be depressed—5.2 percentage points higher. Similarly, the probability of having depression is higher for the age group 41 to 64 years—9.6 percentage points—and for those ages 65 years or more—16.3 percentage points. Men are less likely to have depression—2 percentage points less—than women. Each additional year of education lowers the probability of depression by a modest 0.6 percentage points in absolute terms. Those individuals with a disability have an increased probability of having depression by 14 percentage points. The presence of an illness increases the probability of having depression by 5.7 percentage points. Increasing socioeconomic status, as measured by asset quintiles, indicate those with more access to wealth and resources have lower probability of having depression. As compared to the poorest quintile (Q1), individuals of Q2 were less likely to have depression—2.7 percentage points—with lesser likelihood for members of Q3 at 3.9 percentage points, Q4 at 4.8 percentage points, and Q5 at 6.1 percentage points. A one point increase in the WASH index has a protective effect of lowered depression probability of 1 percentage point. The experience of a household shock—environmental, health, or financial—increases the probability of depression by 4.6 percentage points. Living in a community that disposes of general waste in a designated place increases the probability of depression at 3.3 percentage points. An increased likelihood of depression, at 1.9 percentage points, among those in the middle category of the community connectivity index, compared to the least connected communities, is also observed. Finally, an increase in temperature lowers the probability of depression by 1.6 percent.

Anxiety

Age has a similar effect on anxiety as on depression, with older individuals more vulnerable to having anxiety. Compared to those of ages 15 to 25 years, an increase of 2.1 percentage points exists in the probability of being anxious for those of ages 26 to 40 years, an increase of 1.9 percentage points for the age group 41 to 64 years, and an increase of 4.1 percentage points for those ages 65 years or more (table 4.4). Men are more likely, by 1.1 percentage points, to have anxiety than women. Urban residents are more likely to be anxious at 2.2 percentage points compared to their rural counterparts. Having a disability increases the likelihood of having anxiety by 3.4 percentage points. Similar to depression, increasing socioeconomic status has a protective effect against the probability of having anxiety: Q2, 2 percentage points; Q3, 2.1 percentage points; Q4, 2.5 percentage points; and Q5, 3 percentage points compared to the base, Q1. Experiencing a shock makes it more likely for someone to be anxious by 2.3 percentage points. Living in a community with a designated general waste disposal location increases the probability of anxiety by 3.5 percentage points, while living in a high-connectivity location decreases its likelihood by 3.2 percentage points. Increases in mean humidity and mean temperature increase the probability of having anxiety by 0.3 percent and 0.8 percent, respectively.

REFERENCES

Allen, J., R. Balfour, R. Bell, M. Marmot. 2014. "Social Determinants of Mental Health." *International Review of Psychiatry* 26 (4): 392–407. doi:10.3109/09540261.2014.928270.

Bangladesh NIPORT (National Institute of Population Research and Training) and ICF. 2020. *Bangladesh Demographic and Health Survey 2017–18*. Dhaka: NIPORT and ICF. https://www.dhsprogram.com/pubs/pdf/FR344/FR344.pdf.

Engel, G. L. 1977. "The Need for a New Medical Model: A Challenge for Biomedicine." *Science* 196 (4286): 129. doi:10.1126/science.847460.

Lowe, R., A. M. Stewart-Ibarra, D. Petrova, M. Garcia-Díez, M. J. Borbor-Cordova, R. Mejía, M. Regato, and X. Radó. 2017. "Climate Services for Health: Predicting the Evolution of the 2016 Dengue Season in Machala, Ecuador." *Lancet Planet Health* 1: e142–e151.

Rahman, M. M., S. Ahmad, A. S. Mahmud, M. Hassan-uz-Zaman, M. A. Nahian, A. Ahmed, Q. Nahar, and P. K. Streatfield. 2019. "Health Consequences of Climate Change in Bangladesh: An Overview of the Evidence, Knowledge Gaps and Challenges." *WIREs Climate Change* 10 (5): e601. doi.org/10.1002/wcc.601.

Raza, W. A., E. Van de Poel, P. Panda, D. M. Dror, and A. Bedi. 2015. "Healthcare Seeking Behaviour among Self-Help Group Households in Rural Bihar and Uttar Pradesh, India." *BMC Health Services Research* 16 (1). doi:10.1186/s12913-015-1254-9.

United States, NWS (National Weather Service). 2020. "Meteorological Conversions and Calculation: Heat Index Calculator." Update July 3, 2020. College Park, MD: NWS Weather Prediction Center. https://www.wpc.ncep.noaa.gov/html/heatindex.shtml.

Wu, J., M. Yunus, M. Ali, and M. E. Escamilla. 2018. "Influences of Heatwave, Rainfall, and Tree Cover on Cholera in Bangladesh." *Environment International* 120 (November): 304–11.

III Climate Change and Disease Patterns

5 Change in Climate Patterns: 1976–2019

INTRODUCTION

In this chapter, climate data for Bangladesh is analyzed for the period 1976 to 2019 based on data available from the Bangladesh Meteorological Department (BMD). In addition to minimum and maximum temperatures, information on rainfall and relative humidity were also collected from BMD. This chapter summarizes the patterns of climate change observed for Bangladesh over the past 44 years and links them to the climate suitability for mosquitoes, keeping in view the correlation between weather variables and incidence of dengue, as described in chapter 2. The analyses presented in this chapter's sections reveal that the weather patterns for Dhaka city are quite different from those for Chattogram city. With falling humidity levels, rising temperatures, and increasing rainfall in the summer months, the risk of dengue spread may be higher in Dhaka. These are generalized trends, and more localized information is needed to understand these patterns better.

CHANGES IN TEMPERATURE

Minimum and maximum temperature were analyzed for the entire country using national averages. Since urbanization is one of the key factors that influences the spread of dengue, weather variables for Dhaka and Chattogram cities are separately analyzed.

Bangladesh (national averages)

Bangladesh has gotten warmer over the past 44 years (figure 5.1), with an increase in annual mean temperature by 0.5°C between 1980 and 2019, based on three-year averages.[1]

Figure 5.2 provides a more detailed representation of the pattern of maximum temperature recorded on a monthly basis between 1976 and 2019.

FIGURE 5.1

Annual mean temperature for Bangladesh, 1976–2019

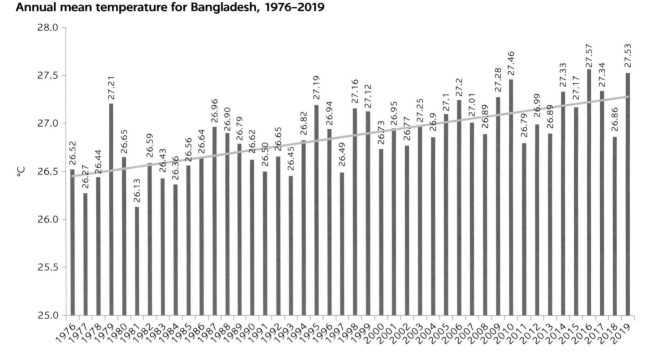

Source: Original figure for this publication.
Note: Bars indicate year-specific mean temperature, and the light blue line represents linear trend.

FIGURE 5.2

National monthly maximum temperature, Bangladesh, 1976–2019

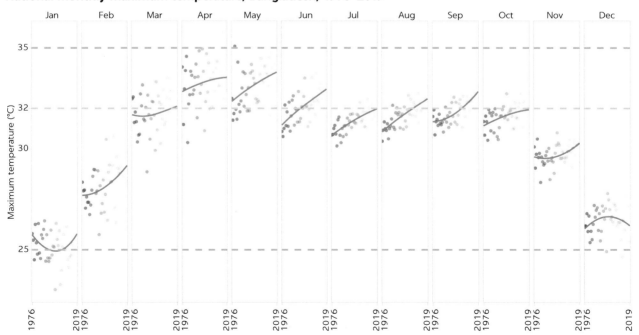

Source: Original figure for this publication.
Note: The figure shows monthly average maximum temperature for every year between 1976 and 2019. The trend lines across the years for each month (in red) are represented by a fitted Lowess curve, which is based on the scatterplot for each month. Horizontal dotted lines at 25°C, 32°C, and 35°C indicate the range of maximum temperature that is conducive to the spread of dengue cases. Between 25°C and 35°C (orange line) the spread increases with a peak at 32°C (blue line). At temperatures higher than 35°C, mosquitoes cannot survive.

The trends suggest that the maximum temperature continued to rise between February and November. As described in chapter 2, dengue cases increase between the maximum temperature range of 25°C to 35°C with a peak at 32°C (dotted horizontal lines in figure 5.2). At high temperatures of 35°C or more, mosquito egg production, the number of eggs hatched, and the life span of female mosquitos decrease significantly. Figure 5.3 analyzes the change in minimum temperature based on monthly averages over the period 1976 to 2019. Evidence indicates minimum temperatures between 18°C and 25°C are conducive to the spread of dengue fever. These thresholds are marked in horizontal dotted lines in figure 5.3.

The temperature in Bangladesh falls well within the range for maximum temperature of 25–35°C conducive to mosquitoes, with summer temperatures of March to July reaching or rising above the 32°C mark. With respect to minimum temperature, a mixed pattern prevails with minimum temperatures for December and February falling close to the 18°C, which is conducive for mosquitoes, while for November and January minimum temperatures show a declining trend. Humidity and precipitation along with suitable temperatures are important for mosquito breeding, and temperature changes alone cannot substantially affect the mosquito life cycle (chapter 2).

Further analysis of maximum temperature by various regions or zones of Bangladesh is presented in map 5.1 using three-year averages.[2] In 1980, the western and central areas of Bangladesh experienced maximum temperature in the range of 30.6°C to 31.0°C, while the southern part ranged from 30.1°C to

FIGURE 5.3

National monthly minimum temperature, Bangladesh, 1976–2019

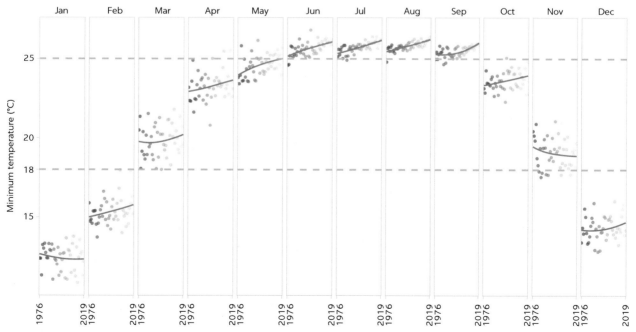

Source: Original figure for this publication.
Note: The figure shows the monthly average minimum temperature recorded for each year between 1976 and 2019. Trend lines across the years for each month (in red) are represented by a fitted Lowess curve, which is based on the scatterplot for each month. Horizontal dotted lines at 18°C and 25°C indicate the minimum temperature range that is suitable for mosquitoes.

30.5°C and eastern areas were the lowest at 29.5°C to 30.0°C. This changed dramatically over a 40-year period. In 2019, the western part of the country recorded higher temperatures in the range of 31.6°C to 32.0°C; the central, southern, and southwest recorded 31.1°C to 31.5°C; and the northeastern part recorded the lowest of all regions, with the range of 30.6°C to 31.0°C. Map 5.2 presents the incremental change in maximum temperature between 1980 and 2019 for these zones or regions using three-year averages as used for map 5.1. While the western part of the country recorded the highest maximum temperature in 2019, the incremental change in temperature over time was the highest for the eastern part of the country, at more than 0.9°C. For the western region, the increase in maximum temperature was in the range of 0.6°C to 0.9°C. It was lowest for the central part, with an increase of less than or equal to 0.5°C.

Dhaka city

Like the rest of the country, the city of Dhaka is becoming warmer, with summers becoming elongated. The maximum temperature in Dhaka has increased by 0.5°C over the past 44 years (map 5.2). Figure 5.4 presents a more detailed analysis of the monthly averages for maximum temperature recorded for Dhaka city between 1976 and 2019. In recent years, average monthly maximum temperature for Dhaka city has risen above 32°C during March to October, with temperatures between April and June reaching close to or above 35°C. As noted earlier, mosquito production declines significantly at high temperatures and can result in a decline in the incidence of dengue. The minimum temperature for

MAP 5.1

Average increase in maximum temperature by Bangladesh zone, 1980 and 2019

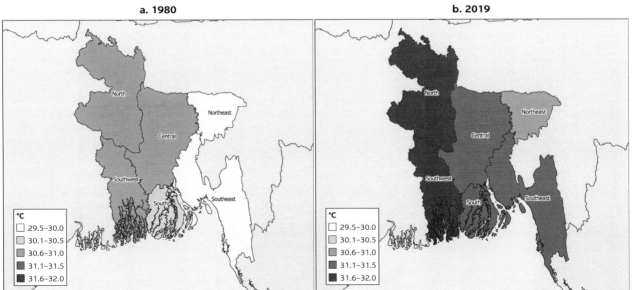

Source: Original map for this publication.
Note: Figures are based on three-year moving averages: that is, 1980 represents average maximum temperature for the years 1978, 1979, and 1980 and 2019 represents an average of 2017, 2018, and 2019. This is done to counter one-off anomalies in weather patterns for any single year over the duration. The zones cover the following administrative divisions of Bangladesh: central, Dhaka and Mymensingh; northeast, Sylhet; southeast, Chattogram; south, Barisal; southwest, Khulna; and north, Rangpur and Rajshahi.

MAP 5.2

Change in maximum temperature for each Bangladesh zone between 1980 and 2019

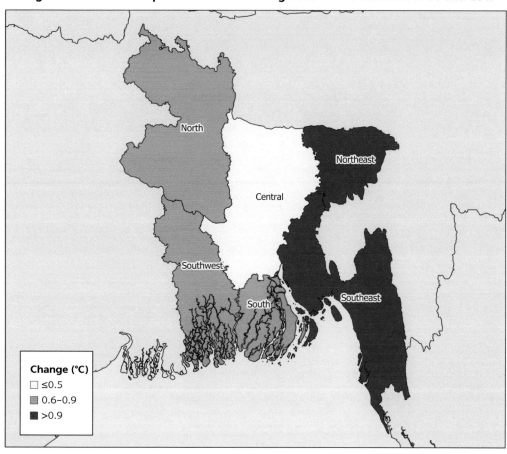

Source: Original map for this publication.
Note: Figures are based on three-year moving averages: that is, 1980 represents average maximum temperature for the years 1978, 1979, and 1980, and 2019 represents an average of 2017, 2018, and 2019. This is done to counter one-off anomalies in weather patterns for any single year over the duration. The zones cover the following administrative divisions of Bangladesh: central, Dhaka and Mymensingh; northeast, Sylhet; southeast, Chattogram; south, Barisal; southwest, Khulna; and north, Rangpur and Rajshahi.

Dhaka city is rising for all months (figure 5.5), which is quite different for the trends presented for the national minimum temperature. In recent years, minimum temperatures for November, December, and February have risen above or reached 18°C, conducive to mosquito breeding.

Chattogram city

Maximum temperature changes for Chattogram city are different from the situation in Dhaka city and the average for all of Bangladesh, with the overall trend decreasing for Chattogram city (figure 5.6). In recent years, the average monthly maximum temperature for Chattogram city has risen above 32°C for the months of March to October but is less than 35°C, unlike Dhaka city.

Patterns of change in minimum temperature for Chattogram city are presented in figure 5.7, which show a rising trend for all months like Dhaka. For the winter months of November to February, minimum temperatures are close

FIGURE 5.4

Maximum monthly average temperature, Dhaka city, 1976–2019

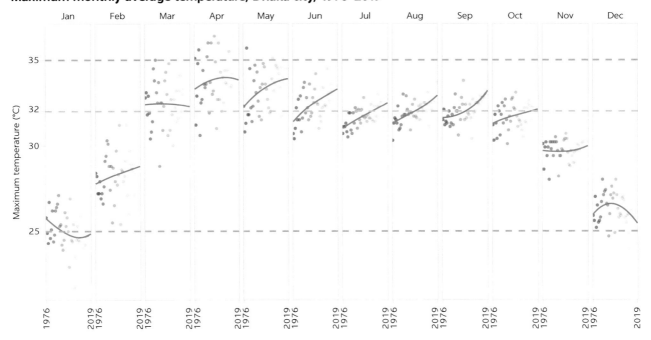

Source: Original figure for this publication.
Note: The figure shows monthly average maximum temperature recorded for Dhaka city for each year between 1976 and 2019. The trend line across the years for each month (in red) is represented by a fitted Lowess curve, which is based on the scatterplot for each month. Horizontal dotted lines at 25°C, 32°C and 35°C indicate the range of maximum temperature that is conducive to the spread of dengue cases; between 25°C and 35°C (orange lines) the spread increases with a peak at 32°C (blue line). At temperatures higher than 35°C, mosquitoes cannot survive.

FIGURE 5.5

Minimum monthly average temperature, Dhaka city, 1976–2019

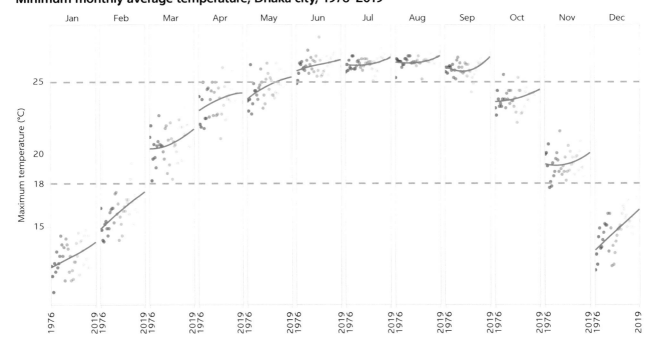

Source: Original figure for this publication.
Note: The figure shows monthly average minimum temperature recorded for Dhaka city for each year between 1976 and 2019. The trend line across the years for each month (in red) is represented by a fitted Lowess curve, which is based on the scatterplot for each month. Horizontal dotted lines at 18°C and 25°C indicate the minimum temperature range that is suitable for mosquitoes.

to or above the 18°C mark, which is conducive for mosquitoes. Humidity and precipitation along with temperatures are important factors for mosquito breeding.

CHANGES IN PRECIPITATION

In this section, precipitation data recorded by BMD between 1976 and 2019 are analyzed. Rainfall-related data are very location specific, and using national averages for analysis is not the best possible option. The 43 weather stations in Bangladesh are not uniformly distributed across the country. In the absence of more reliable data, this generalized analysis is being used to estimate overall trends.

Figure 5.8 shows changes in annual mean rainfall for Bangladesh for the years 1976 to 2019, using three-year averages.[3] The variation in rainfall across the years is quite substantial, which is why the trend line represented by the horizonal blue line in figure 5.8 shows no major change. Overall, the rainfall patterns in Bangladesh have been in the range that is conducive to mosquitoes: 200 to 800 millimeters of rain.

Map 5.3 shows the variation in rainfall patterns in various parts of the country for the years 1980 and 2019, using three-year averages. Over time, rainfall recorded for the southern part of Bangladesh appears to have decreased: in 1980, average rainfall recorded for southwest and south was in the range of 150–200 millimeters, which declined to 129–150 millimeters.

FIGURE 5.6

Maximum monthly average temperature, Chattogram city, 1976–2019

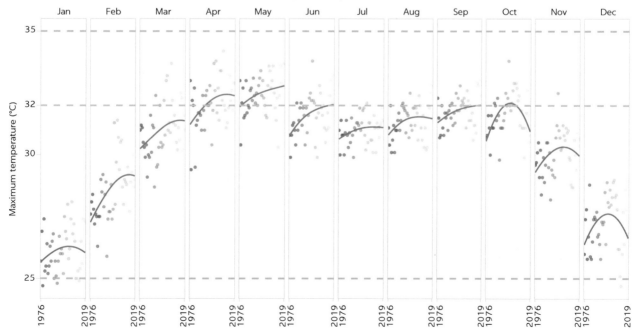

Source: Original figure for this publication.
Note: The figure shows monthly average maximum temperature recorded for Chattogram city for each year between 1976 and 2019. The trend line across the years for each month (in red) is represented by a fitted Lowess curve, which is based on the scatterplot for each month. Horizontal dotted lines at 25°C, 32°C, and 35°C indicate the range of maximum temperature that is conducive to the spread of dengue cases; between 25°C and 35°C (orange lines) the spread increases with a peak at 32°C (blue line). At temperatures higher than 35°C, mosquitoes cannot survive.

FIGURE 5.7

Minimum monthly average temperature, Chattogram city, 1976–2019

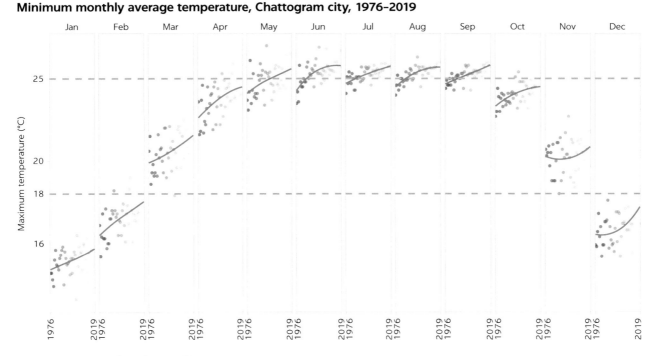

Source: Original figure for this publication.
Note: The figure shows monthly average minimum temperature recorded for Chattogram city for each year between 1976 and 2019. The trend line across the years for each month (in red) is represented by a fitted Lowess curve, which is based on the scatterplot for each month. Horizontal dotted lines at 18°C and 25°C indicate the minimum temperature range that is suitable for mosquitoes.

FIGURE 5.8

Annual mean rainfall for Bangladesh, 1976–2019

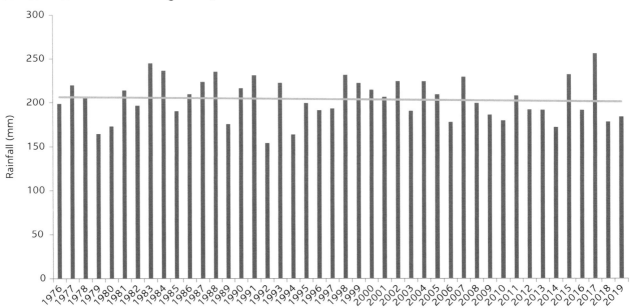

Source: Original figure for this publication.
Note: The horizontal blue line represents the trend while the vertical bars indicate average rainfall in millimeters for the specific years indicated.

MAP 5.3

Rainfall patterns across Bangladesh, 1980 and 2019

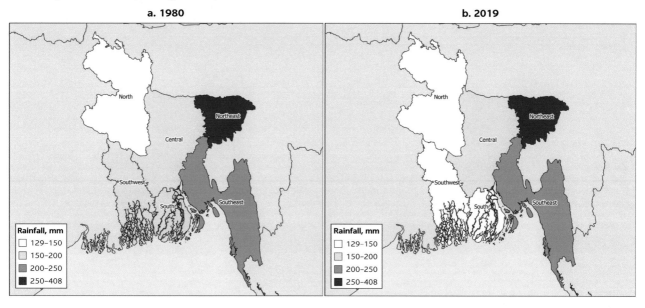

a. 1980

b. 2019

Source: Original map for this publication.
Note: Figures are based on three-year moving averages: that is, 1980 represents average rainfall for the years 1978, 1979, and 1980, and 2019 represents an average of 2017, 2018, and 2019. This is done to counter one-off anomalies in weather patterns for any single year over the duration. The zones cover the following administrative divisions of Bangladesh: central, Dhaka and Mymensingh; northeast, Sylhet; southeast, Chattogram; south, Barisal; southwest, Khulna; and north, Rangpur and Rajshahi.

With this change, the western and southern part of Bangladesh recorded the lowest average rainfall in 2019. As mentioned, this is a generalized analysis to document overall patterns, in the absence of more localized information. Further analysis of rainfall data for Dhaka and Chattogram cities separately follow.

Rainfall patterns for the cities of Dhaka and Chattogram are quite different, with Chattogram experiencing higher rainfall than Dhaka over the past 44 years, on average (figures 5.9 and 5.10). Average rainfall for Dhaka city has been increasing between the months of April and August, except in May, when the overall trend indicates a decreasing pattern. More important, average rainfall is within the range of 200 to 800 millimeters, which is conducive to mosquito breeding, as discussed in chapter 2. Average monthly rainfall in Chattogram city for the months of June to August was close to the 800-millimeter mark in recent years, and trends for June and July indicate it may increase further. Based only on rainfall patterns for the two cities, Dhaka appears to be more conducive to mosquito breeding than Chattogram. Rainfall along with humidity and temperature determine climate suitability for mosquitoes. It should also be noted that rainfall data for Dhaka and Chattogram as presented are not very accurate, because both of these divisions are geographically spread out over large areas, and one weather station each for Dhaka and Chattogram cities is insufficient. More localized information is required for accurate analysis and predictions.

FIGURE 5.9

Monthly average rainfall for Dhaka city, 1976–2019

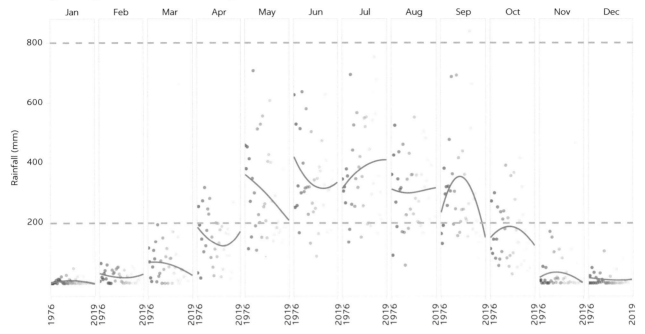

Source: Original figure for this publication.
Note: The figure shows monthly average rainfall recorded for Dhaka city for each year between 1976 and 2019. Horizontal dotted lines at 200 and 800 millimeters indicate a range that is most conducive to mosquito breeding. The trend line across the years for each month (in red) is represented by a fitted Lowess curve, which is based on the scatterplot for each month.

FIGURE 5.10

Monthly average rainfall for Chattogram city, 1976–2019

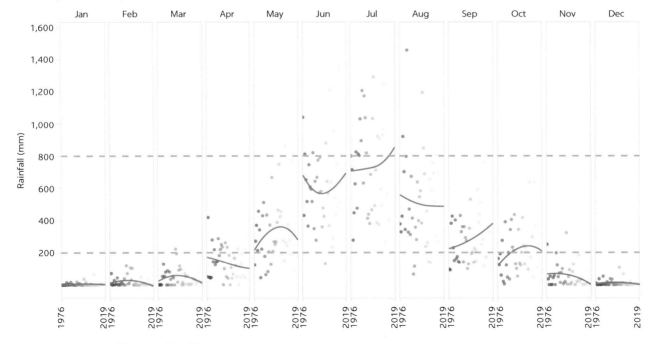

Source: Original figure for this publication.
Note: The figure shows trends in monthly average rainfall recorded for Chattogram city for each year between 1976 and 2019. Horizontal dotted lines at 200 and 800 millimeters indicate a range that is most conducive to mosquito breeding. The trend line across the years for each month (in red) is represented by a fitted Lowess curve, which is based on the scatterplot for each month.

CHANGES IN RELATIVE HUMIDITY

In this section, data on relative humidity recorded by BMD between 1976 and 2019 are analyzed. Like rainfall, humidity-related data are location specific, and using national averages for analysis is not the best possible option. The 43 weather stations in Bangladesh are not uniformly distributed across the country. In the absence of more accurate data, this generalized analysis is used to estimate overall trends.

Map 5.4 shows the relative humidity in various parts of the country and how it has changed between 1980 and 2019, using three-year averages. It appears that the north, northeast, and south of Bangladesh are becoming more humid over time, and in 2019 humidity levels were equal to or more than 80 percent, which is not conducive for mosquitoes.[4]

The pattern of changes in humidity in Dhaka and Chattogram cities is significantly different. In Dhaka, average monthly humidity is declining (figure 5.11), while in Chattogram it is increasing during summer (figure 5.12), especially in June and July. Moreover, average humidity in Dhaka city is within the range of 60 and 80 percent, which is conducive to mosquitoes (as explained in chapter 2). Particularly in recent years, humidity in the summer months in Dhaka is falling and is below 80 percent. In Chattogram city, however, humidity in recent years was close to or higher than 80 percent. For mosquitoes spreading dengue, lower levels of humidity (with high temperature) are more conducive to breeding and reproduction.

MAP 5.4

Variation in relative humidity across Bangladesh, 1980 and 2019

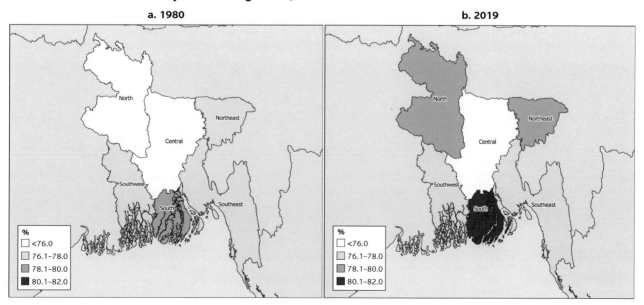

Source: Original map for this publication.
Note: Figures are based on three-year moving averages: that is, 1980 represents average humidity for the years 1978, 1979, and 1980 and 2019 represents an average of 2017, 2018, and 2019. The zones cover the following administrative divisions of Bangladesh: central, Dhaka and Mymensingh; northeast, Sylhet; southeast, Chattogram; south, Barisal; southwest, Khulna; and north, Rangpur and Rajshahi.

FIGURE 5.11

Humidity for Dhaka city, 1976–2019

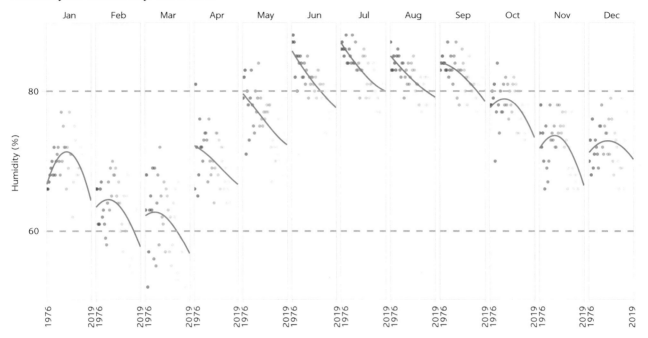

Source: Original figure for this publication.
Note: The figure shows monthly average humidity recorded for Dhaka city for each year between 1976 and 2019. Horizontal dotted lines at 60 percent and 80 percent represent the range most conducive to mosquito breeding. The trend line across the years for each month (in red) is represented by a fitted Lowess curve, which is based on the scatterplot for each month.

FIGURE 5.12

Humidity for Chattogram city, 1976–2019

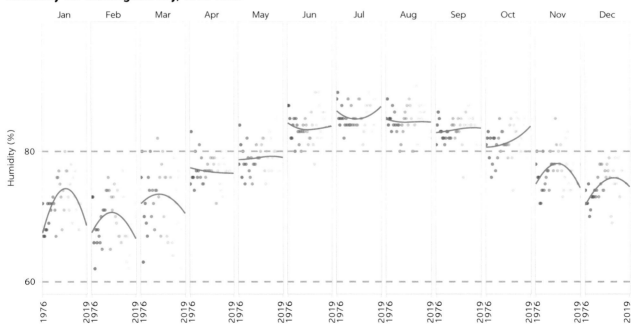

Source: Original figure for this publication.
Note: The figure shows monthly average humidity recorded for Chattogram city for each year between 1976 and 2019. Horizontal dotted lines at 60 percent and 80 percent represent the range most conducive to mosquito breeding. The trend line across the years for each month (in red) is represented by a fitted Lowess curve, which is based on the scatterplot for each month.

HEAT INDEX

Heat waves and exposure to high temperatures have negative impacts on human health, with morbidity and mortality caused by heat-related stress (Watts et al. 2020). Densely populated areas of the world are increasingly exposed to warmer climatic conditions, experiencing higher change in mean summer temperature compared to the global average (WMO 2020). Such extreme heat conditions are taking a toll on human health and overwhelming the health systems, with greater consequences for places where extreme heat occurs in the context of aging populations, urbanization, urban heat island effects, and health inequalities (WMO 2020). The elderly, people with disabilities or preexisting medical conditions or with both, and those exposed to heat from working outdoors or in noncooled environments are the worst affected by heat waves (Watts et al. 2020). Heat-related mortality among the elderly population—people more than 65 years of age—has increased by 53.7 percent in the past 20 years (Watts et al. 2020). Due to heat stress, people's productivity at work is impacted, and in Bangladesh an estimated 148 work hours per person was lost in 2019, which translates to 18.2 billion work hours lost in total for the country, compared to 13.3 billion work hours lost in 2013 (Watts et al. 2020).

The change in heat index expressed in degrees Celsius for Dhaka and Chattogram cities is analyzed for the period 1976 to 2019 (figure 5.13) to provide an overview of how heat affected these cities. Heat index is a measure of real feel when relative humidity is factored in with the actual air temperature (United States, NWS 2020). The heat index was constructed using BMD's data for maximum temperature and humidity to represent composite conditions using the Rothfusz (1990) equation. Overall, the heat indexes indicate a "danger" level during the months of April to October, with little variation over the years. A danger level of the heat index indicates that heat cramps and heat exhaustion are likely while heat stroke is probable with continued outdoor activity. The major difference between Dhaka and Chattogram cities is noted in January, with Dhaka being relatively cooler than Chattogram, shown by more light blue cells in figure 5.13 (indicating a temperature range that will not adversely affect humans, with heat in normal limits of less than 27°C). A deeper look at the two indexes reveals differences over time: for example, in 1976 there was a range of 40°C and 45°C for Dhaka and 40°C and 47°C for Chattogram; in 2018–19, the range increased to 45°C and 51°C for both the cities.

NOTES

1. This is done to counter one-off anomalies in weather patterns for any single year over the duration. The overall average is based on the average of annual maximum temperatures for the years 1978–80 and for the years 2016–19.
2. This is done to counter one-off anomalies in weather patterns for any single year over the duration. The overall average is based on the average of annual maximum temperatures for the years 1978–80 and for the years 2016–19.
3. This is done to counter one-off anomalies in weather patterns for any single year over the duration. The overall average is based on the average of annual maximum temperatures for the years 1978–80 and for the years 2016–19.
4. As mentioned in chapter 2, relative humidity levels of 60–80 percent is conducive to mosquito breeding.

FIGURE 5.13

Heat index for cities of Dhaka and Chattogram for each month between 1976 and 2019
°C

Dhaka	1	2	3	4	5	6	7	8	9	10	11	12		Chattogram	1	2	3	4	5	6	7	8	9	10	11	12
1976	27	31	40	38	45	42	42	41	44	41	36	27		1976	27	30	40	46	47	40		43	46	40	36	27
1977	25	31	43	42	41	42	44	43	47	38	33	27		1977	28	31	42	37	40	39	40	42	44	41	37	30
1978	25	28	35	44	43	44	44	44	43	44	35	29		1978	25	31	36	43	45	42	44	46	42	44	37	31
1979	28	28	38	47	54	46	44	45	44	42	37	26		1979	29	29	36	43	45	42	42	40	42	41	38	27
1980	25	29	38	51	44	45	42	43	43	39	34	28		1980	26	29	36	44	43	43	42	41	46	38	33	28
1981	26	28	35	38	43	47	42	47	45	41	34	26		1981	26	30	35	36	42	42	40	43	42	41	36	
1982	28	29	35	43	50	45	46	42	46	43	31	27		1982	28	30	35	41	46	41	42	40	41	41	32	27
1983	25	28	38	43	45	50	46	43	44	41	35	27		1983	26	28	34	41	43	45	43	42	44	42	35	27
1984	25	29	42	48	43	42	43	43	43	43	34	28		1984	26	29	36	42	45	43	42	43	45	46	36	28
1985	28	31	46	47	45	45	42	45	45	44	36	31		1985	28	30	37	44	44	44	40	44	44	43	33	30
1986	28	31	43	47	49	51	46	49	43	42	35	30		1986	27	32	39	41	45	45	42	45	42	43	35	30
1987	28	34	38	47	52	52	45	45	46	44	36	30		1987	28	32	36	41	45	45	42	42	45	45	38	30
1988	29	35	41	53	49	46	47	45	49	46	37	30		1988	29	33	39	44	45	43	44	44	45	44	38	31
1989	26	31	41	52	52	48	46	47	47	43	36	28		1989	25	31	37	42	46	44	42	44	43	38	35	29
1990	26	29	33	42	42	46	44	46	46	37	36	28		1990	27	32	33	40	45	44	39	45	46	41	38	29
1991	25	31	40	48	45	45	46	45	42	41	31	25		1991	26	33	43		44	40	43	46	43	42	32	28
1992	24	26	39	56	47	50	44	44	45	43	34	27		1992	25	27	37	45	45	45	43	45	45	42	35	27
1993	25	32	35	43	42	44	43	42	44	43	35	29		1993	27	30	35	44	45	42	43	42	43	43	35	30
1994	28	28	40	44	48	44	44	44	46	45	36	30		1994	29	30	37	43	47	45	44	46	47	44	37	30
1995	26	30	40	52	53	48	45	47	47	47	36	29		1995	27	30	39	46	48	47	43	45	45	47	36	30
1996	27	33	46	49	53	47	49	45	53	44	37	29		1996	28	32	39	44	49	47	46	42	47	45	40	33
1997	25	29	41	39	48	46	45	48	45	43	37	26		1997	27	30	39	38	46	46	44	48	46	45	41	28
1998	23	31	35	44	48	53	46	45	48	49	39	32		1998	26	31	36	43	49	53	45	45	49	52	46	34
1999	29	36	43	53	46	47	44	44	44	44	37	29		1999	31	38	43	50	46	48	46	44	46	46	40	32
2000	25	26	35	44	44	45	43	44	44	41	35	28		2000	29	31	39	46	45	47	46	46	47	46	40	31
2001	25	31	37	46	43	42	43	46	45	43	34	27		2001	28	35	42	47	42	39	41	44	45	44	37	32
2002	26	30	36	41	44	45	45	43	45	40	33	27		2002	31	38	42	40	41	42	39	42	44	44	37	31
2003	22	30	34	47	47	43	45	45	44	42	33	27		2003	27	36	39	46	47	40	47	46	46	50	39	32
2004	24	30	39	43	49	46	43	44	41	39	33	29		2004	28	33	40	42	50	45	42	45	44	46	39	33
2005	25	31	40	46	46	49	43	45	48	39	33	29		2005	29	35	39	47	48	48	44	43	50	51	40	35
2006	26	37	38	44	47	46	46	45	44	43	34	28		2006	32	39	44	47	48	51	45	46	47	48	41	32
2007	25	29	34	45	49	46	44	46	44	42	34	27		2007	29	33	39	45	49	47	43	47	45	47	42	32
2008	25	31	38	45	49	46	45	45	47	41	34	27		2008	30	30	42	49	49	44	42	44	47	46	39	31
2009	27	31	38	50	50	51	46	47	47	42	34	27		2009	28	32	40	44	46	45	40	42	44	42	36	28
2010	24	30	42	50	48	48	47	48	46	42	35	27		2010	25	31	39	43	44	44	44	45	45	45	36	28
2011	24	30	36	42	48	47	45	42	45	42	34	25		2011	25	32	36	40	42	41	43	41	41	41	33	26
2012	24	29	39	45	49	47	46	45	47	42	32	24		2012	26	32	37	42	46	43	41	44	43	43	33	25
2013	24	30	39	44	43	48	45	44	47	42	34	28		2013	25	32	40	43	40	45	43	40	43	39	34	28
2014	25	28	36	47	50	48	46	46	46	42	35	25		2014	26	30	36	47	45	43	43	41	42	41	37	30
2015	25	30	35	42	47	46	44	46	46	43	35	27		2015	28	32	39	44	49	47	43	46	48	43	36	28
2016	25	33	39	52	49	47	46	47	48	45	35	30		2016	27	34	42	46	47	48	43	45	48	46	35	33
2017	27	31	35	43	50	48	45	48	49	44	35	29		2017	29	34	36	43	49	48	45	48	46	43	39	31
2018	24	32	40	43	45	51	48	50	51	41	35	27		2018	25	32	41	46	47	47	46	48	49	39	37	29
2019	28	30	36	45	50	49	49	50	49	44				2019	29	33	38	45	50	49	46	44	48	44		

27–32°C	Caution	Fatigue is possible with prolonged exposure and activity. Continuing activity could result in heat cramps.
33–40°C	Extreme caution	Heat cramps and heat exhaustion are possible. Continuing activity could result in heat stroke.
41–54°C	Danger	Heat cramps and heat exhaustion are likely; heat stroke is probable with continued activity.
Over 54°C	Extreme danger	Heat stroke is imminent.

Source: Original figure for this publication.
Note: Heat index is a measure of "real feel" that combines relative humidity and actual air temperature (United States, NWS 2020). Empty cells indicate no data were available.

REFERENCES

Rothfusz, L. P. 1990. "The Heat Index 'Equation' (or, More Than You Ever Want to Know about Heat Index)." Technical Attachment SR 90-23, July 1, 1990. Fort Worth, TX: National Weather Service Southern Region Headquarters. https://www.weather.gov /media/ffc/ta_htindx.PDF.

United States, NWS (National Weather Service). 2020. "Meteorological Conversions and Calculation: Heat Index Calculator." Update July 3, 2020. College Park, MD: NWS Weather Prediction Center. https://www.wpc.ncep.noaa.gov/html/heatindex.shtml.

Watts, N., M. Amann, N. Arnell, S. Ayeb-Karlsson, L. Beagley, K. Belesova, M. Boykoff, et al. 2020. *Responding to Convergence Crises.* The 2020 report of the *Lancet* Countdown on health and climate change. *Lancet* 397 (10269): 129–70. doi:10.1016/S0140-6736(20)32290-X.

WMO (World Meteorological Organization). 2020. *WMO Statement on the State of the Global Climate in 2019.* WMO no. 1248. Geneva: WMO. https://public.wmo.int/en/resources /library/wmo-statement-state-of-global-climate-2019.

6 Climate Change Observed: 1901–2019

The erratic nature of rainfall and temperature has increased in Bangladesh. Significant increasing trends in cyclone frequency have been observed during November and May, the main months for cyclone activity in the Bay of Bengal. This chapter focuses on changes in annual temperature and precipitation for the past 100 and more years.

Bangladesh has a humid, warm climate influenced primarily by monsoon and partly by premonsoon and postmonsoon circulations. Average temperatures approximate 26.1°C but can vary between 15°C and 34°C throughout the year. The warmest months coincide with the rainy season—March to September—while winter from December to February receives less rainfall. Bangladesh is one of the wettest countries of the world, with most areas receiving at least 1,500 millimeters of rainfall per year and can experience as much as 5,800 millimeters. Rainfall is driven by the southwest monsoon, which originates over the Indian Ocean and carries warm, moist, and unstable air, beginning approximately during the first week of June and ending in the first week of October. Major climate drivers in Bangladesh include the easterly trade winds, the southwest monsoon, and the El Niño Southern Oscillation.

Figure 6.1 shows the change in mean monthly temperature between 1901 and 2019 using averages for two 30-year periods, 1901–30 and 1991–2019. Overall, annual mean temperatures have increased. Summers are hotter and longer—temperatures for March to October rising by 1.1°C to 1.3°C except for July and September, when it increased by 0.8°C. Winters are also becoming warmer, with average annual temperature rising by 1.6°C to 1.8°C in November and December; in January it increased by 0.6°C. The maximum change in annual temperature is seen in February, when it increased by almost 1.9°C.

Changes in mean monthly precipitation between 1901 and 2019 using averages for two 30-year periods 1901–30 and 1991–2019 are presented in figure 6.2. For the peak monsoon season from June to August, average monthly mean rainfall has declined by about 60 millimeters, except for July when it declined by 1.34 millimeters. Mean monthly rainfall for September and October has increased by about 43 millimeters, which indicates the monsoon period is becoming longer, extending from February to October.

FIGURE 6.1

Change in historical mean monthly temperature, 1901–2019

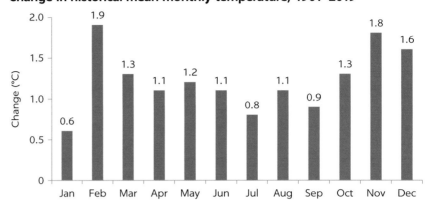

Source: Calculation based on data from the Bangladesh Meteorological Department and the Climate Change Knowledge Portal database of the World Bank, https://climateknowledgeportal.worldbank.org.
Note: The difference is based on two 30-year averages: 1901 to 1930 and 1991 to 2019.

FIGURE 6.2

Change in historical mean monthly rainfall, 1901–2019

Source: Calculation based on data from the Bangladesh Meteorological Department and the Climate Change Knowledge Portal database of the World Bank, https://climateknowledgeportal.worldbank.org.
Note: The difference is based on two 30-year averages: 1901 to 1930 and 1991 to 2019.

Although Bangladesh is said to have six seasons according to the Bengali calendar year, marked by distinct weather features, the analyses in this section indicate these distinctions are getting blurred. Summers are becoming hotter and longer, now spanning from February to October, while the monsoon is also spread over a longer period between February and October with the peak monsoon experiencing less rainfall. Winters are becoming warmer. In essence, Bangladesh appears to be losing its seasonality.

7 Climate Projections to 2099

Using data from the World Bank Climate Change Knowledge Portal, this chapter discusses climate projections up to 2099.[1] According to an assessment conducted by the Intergovernmental Panel on Climate Change (IPCC), continued emissions of greenhouse gases will cause further warming in Bangladesh (IPCC 2014). Mean temperatures across Bangladesh are projected to increase by 1.4°C and 2.4°C by 2050 and 2100, respectively. This warming is expected to be more pronounced in the winter months from December to February. Map 7.1 presents the projected temperature changes for Bangladesh. Observed data indicate that the temperature is generally increasing in the monsoon season of June to August. Average monsoon season maximum and minimum temperatures show an increasing trend annually at the rate of 0.05°C and 0.03°C, respectively.

Rising temperatures leading to more intense and unpredictable rainfalls during the monsoon season and a higher probability of catastrophic cyclones are expected to result in increased tidal inundation. This was evident during the recent super-cyclone Amphan in 2020, when Bangladesh experienced heavy monsoon rains that led to flooding in various parts of the country, whereby greater than one-third of the country was flooded, with more than 4.9 million people affected and 42 lives lost.

Map 7.2 presents the projected change in maximum rainfall in Bangladesh. Peak five-day rainfall intensity—a substitute for an extreme storm event—is projected to increase. Annual precipitation will rise by 74 millimeters by 2040–59. The frequency of tropical cyclones in the Bay of Bengal may increase, and according to the IPCC (2001) there is evidence that the peak intensity may increase by 5 percent to 10 percent and precipitation rates may increase by 20 percent to 30 percent. Cyclone-induced storm surges are likely to be exacerbated by a potential rise in sea level of more than 27 centimeters by 2050.

MAP 7.1

Projected change in monthly temperature compared to 1986–2005

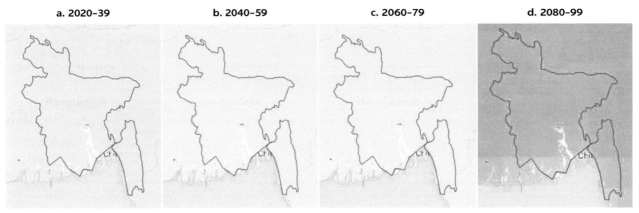

a. 2020–39   b. 2040–59   c. 2060–79   d. 2080–99

Scale

Monthly temperature (°C)

−5   −4   −3   −2   −1   0   1   2   3   4   5   6

Source: World Bank Climate Change Knowledge Portal database.
Note: Future climate information is derived from 35 available global circulation models used by the IPCC (2014) fifth assessment report. Data are presented at a 1° x 1° global grid spacing, produced through bilinear interpolation.

MAP 7.2

Projected change in maximum five-day rainfall compared to 1986–2005

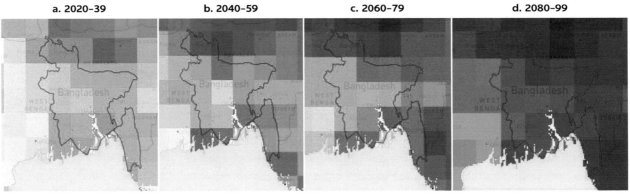

a. 2020–39   b. 2040–59   c. 2060–79   d. 2080–99

Scale

Maximum five-day rainfall (mm)

−60   −50   −40   −30   −20   −10   0   10   20   30   60   90   120

Source: Climate Change Knowledge Portal of the World Bank database.
Note: Change in the largest consecutive five-day cumulative precipitation sum per month or year relative to the reference period (1986–2005). These values are often larger in the warm season, and they are broadly expected to increase as the atmosphere has a higher capacity to carry moisture along as temperatures warm.

NOTE

1. Climate projections are available from the World Bank Climate Change Knowledge Portal database, "Country: Bangladesh," https://climateknowledgeportal.worldbank.org/country /bangladesh/climate-data-projections.

REFERENCES

IPCC (Intergovernmental Panel on Climate Change). 2001. *Climate Change 2001: Synthesis Report: A Contribution of Working Groups I, II, and III to the Third Assessment Report of the Intergovernmental Panel on Climate Change.* Edited by R. T. Watson and the Core Writing Team. Cambridge, UK: Cambridge University Press.

IPCC (Intergovernmental Panel on Climate Change). 2014. *Climate Change 2014: Synthesis Report: Contribution of Working Groups I, II, and III to the Fifth Assessment Report of the Intergovernmental Panel on Climate Change.* Edited by R. K. Pachauri, L. A. Meyer, and the Core Writing Team. Geneva: IPCC.

8 Patterns of Infectious Diseases in Bangladesh

INTRODUCTION

In this chapter, data for vector- and waterborne diseases are analyzed in line with the evidence presented in chapter 1. The incidence and mortality related to dengue for Bangladesh are showing an increasing trend, but these are declining for diarrhea and malaria. This finding aligns with existing literature, which shows incidence of dengue is high in many regions where malaria has been effectively controlled or eradicated (Ooi et al. 2010).

PATTERNS OF DENGUE, 2000–20

The first outbreak of dengue in Bangladesh was recorded in 1964 and was termed "Dacca fever" at that time. Sporadic outbreaks of dengue followed between 1964 and 1999 but were not officially recorded (Hsan et al. 2019). The first official outbreak of dengue was reported in 2000 with 5,551 cases and 93 deaths, and 2019 recorded the highest incidence of 101,354 cases and 164 deaths (figure 8.1). These figures are likely to be underreported, as national surveillance is passive (Mamun et al. 2019) and only cases reported in hospitals are included.

Table 8.1 summarizes the incidence and deaths caused by dengue by location. Half of the total dengue cases reported in 2019 for Bangladesh was in Dhaka city alone, with 51 percent of the cases, while the entire Dhaka division, including Dhaka city, reported 63 percent of the total cases. Most of the dengue-related deaths were also in Dhaka division, at 82 percent, with Dhaka city accounting for 77 percent of the total deaths reported in the country. After Dhaka, Khulna division recorded the second highest incidence at 12 percent and a dengue-related mortality of 9 percent of the total recorded for the country.

Dengue is usually prevalent between July and September in Bangladesh. Based on data between 2000 and 2017, almost 50 percent of the dengue cases occurred during the monsoon season and 49 percent during the postmonsoon season (Mutsuddy et al. 2018). Since 2014, however, these trends have been changing, and between 2015 and 2017, dengue cases were reported to be more than seven times higher in the premonsoon season, compared to the previous

FIGURE 8.1

Dengue incidence and deaths in Bangladesh, 2000–20

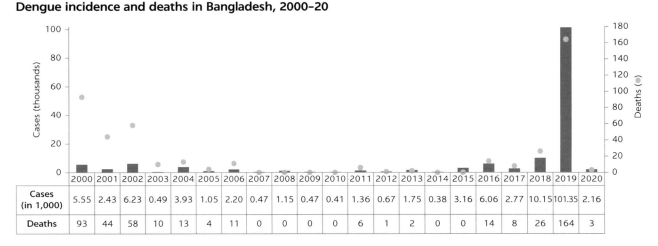

Cases (in 1,000)	5.55	2.43	6.23	0.49	3.93	1.05	2.20	0.47	1.15	0.47	0.41	1.36	0.67	1.75	0.38	3.16	6.06	2.77	10.15	101.35	2.16
Deaths	93	44	58	10	13	4	11	0	0	0	0	6	1	2	0	0	14	8	26	164	3

Sources: Mamun et al. 2019; the Bangladesh Directorate General of Health Services (DGHS) Health Emergency Operations Center for 2019 and 2020.

TABLE 8.1 Dengue fever cases and deaths in Bangladesh, 2019

LOCATION	CASES		DEATHS	
	NUMBER	% OF TOTAL	NUMBER	% OF TOTAL
Dhaka city	51,810	51	127	77
Dhaka division (excluding Dhaka city)	11,743	12	8	5
Dhaka division (total)	63,553	63	135	82
Chattogram division	8,359	8	2	1
Khulna division	11,975	12	15	9
Rangpur division	2,175	2	2	1
Rajshahi division	4,814	5	2	1
Barisal division	7,086	7	5	3
Sylhet division	1,011	1	1	1
Mymensingh division	2,381	2	2	1
Bangladesh				
(Total rounded to nearest whole number)	**101,354**	**100**	**164**	**100**

Source: The Bangladesh Directorate General of Health Services (DGHS) Health Emergency Operations Center.

14 years. Figure 8.2 shows the monthly pattern of dengue for the period 2008 to 2019. In 2019, incidence of dengue was reported for an extended period, with an early onset in June and ending in November.

Climate change or variability, unplanned rapid urbanization, high population densities, and insufficient preparedness, including inadequate public health infrastructure and suboptimal vector-control programs, contribute to the magnitude and severity of dengue outbreaks in Bangladesh (Hsan et al. 2019). The following subsection analyzes the weather variables for Dhaka and Khulna for 2019—the two divisions that recorded the highest disease incidence and dengue-related mortality—to assess if climate variability contributed to the 2019 outbreak in Dhaka.

FIGURE 8.2

Distribution of dengue fever cases by month in Bangladesh, 2008–19

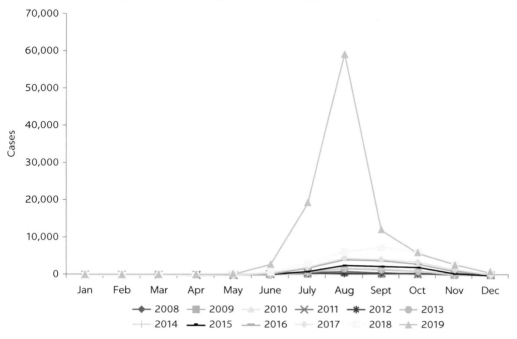

Source: Mamun et al. 2019.

DID WEATHER CONDITIONS CONTRIBUTE TO THE 2019 DENGUE OUTBREAK IN DHAKA?

Yes, perhaps, due to heavy rain in February followed by favorable temperatures and humidity in the subsequent months. As mentioned in chapter 2, conducive weather conditions determine disease transmission; for even if competent species or vectors are present, the pathogen will not spread if temperatures are not suitable.

Dhaka was ranked the sixth most populous city in the world in 2020, with 29,069 people per square kilometer (Wright 2020), and is the most densely populated city in Bangladesh, followed by Chattogram in the second spot and Khulna, the third (WPR 2021). In this section, weather variables for Dhaka in 2019 are analyzed in detail to understand if they contributed to the higher dengue cases and deaths in Dhaka. As mentioned in chapter 2, the risk of dengue fever increases at the onset of rainfall and can last for at least three months. A quick comparison of the data presented in figure 8.3 and figure 8.4 indicates that the climate was more conducive for mosquito breeding in Dhaka than in other areas. Maximum temperature was in the range of 25°C and 35°C for Dhaka and Chattogram, but for Khulna it was close to or exceeded 35°C in the months of April to June. Rainfall was higher in Dhaka and Khulna than in Chattogram for the months of February to June, with the exception of May. With respect to relative humidity, Dhaka was in the range of 60 to 80 percent, which is conducive to breeding mosquitoes, while Khulna and Chattogram were close to or exceeded

FIGURE 8.3

Minimum and maximum temperature and relative humidity for Chattogram, Dhaka, and Khulna, February–July 2019

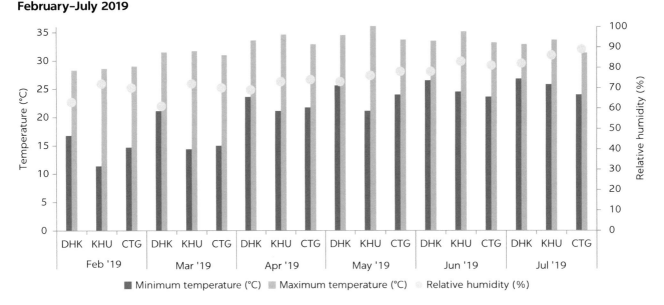

Source: Original figure for this publication.
Note: DHK = Dhaka, KHU = Khulna, and CTG = Chattogram. As noted in earlier sections, climatic conditions suitable for mosquitoes spreading dengue are maximum temperature in the range of 25°C to 35°C, with a peak at 32°C; minimum temperature in the range of 18°C to 25°C; and relative humidity in the range of 60 to 80 percent.

the 80 percent mark between May and July. In July 2019, although the maximum temperature for Chattogram was the lowest among the three areas, rainfall for Chattogram was the highest at 1,135 millimeters—much higher than the conducive range of 200 to 800 millimeters—which probably helped wash away the mosquitoes and their larvae.

Rainfall recorded for the month of February over 44 years is analyzed (figure 8.4). Of the three divisions, Khulna recorded the highest rainfall overall between 1976 and 2020, with 166 millimeters of rain in 2019. The same year, Dhaka recorded the highest rainfall in its history with 115 millimeters. Such high levels of rainfall were previously recorded in Khulna and Chattogram in the 1990s, before 2000 when the dengue outbreak was first officially recorded in Bangladesh. Other factors remaining constant, these weather conditions—temperature, humidity, and rainfall—optimally aided the spread of the pathogen and may help explain the dengue outbreak in Dhaka in 2019.

MALARIA, 2000–20

Malaria is endemic in 91 countries at present representing 3.2 billion people at risk, which is almost half of the world's total population (WHO 2016). The disease is a major public health problem in Bangladesh. It is highly endemic in 13 out of 64 districts, and approximately 11 million people are at risk of the disease (WHO n.d.). In 2008, all these 13 districts were highly endemic areas, but since 2013, only three districts are classified as such (map 8.1). The forested and hilly terrain of Bandarban, Khagrachari, Rangamati, and Cox's Bazar districts have

FIGURE 8.4
February rainfall for Chattogram, Dhaka, and Khulna, 1976–2020

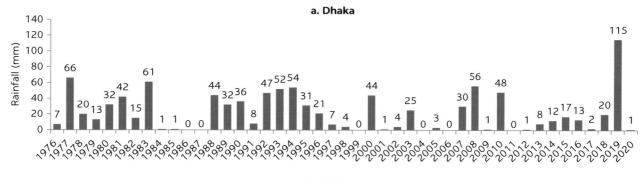

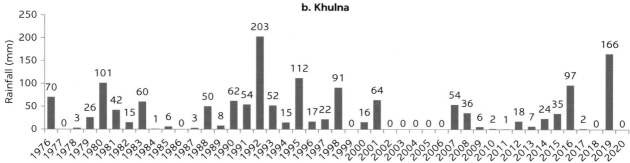

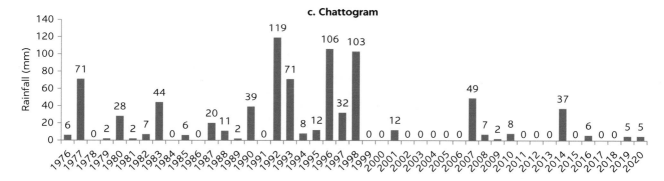

Source: Original figure for this publication.

the geophysical potential for intense malaria transmission throughout the year, and increased mobility of the nonimmune population in the Chattogram Hill Tract districts further adds to the risk of transmission (WHO n.d.). As shown in figure 8.5, the total reported cases of malaria and incidence per 1,000 population at risk declined significantly for Bangladesh between 2000 and 2018. The incidence rate has declined from 7.2 cases per 1,000 population at risk in 2000 to 0.7 in 2018, while the total reported cases was 10,426 in 2018 compared to 55,599 in 2000.

The decline in malaria prevalence is due to a successful intervention by the Ministry of Health and Family Welfare through its National Malaria Control Program. A study based on malaria data between 2008 and 2012 concluded "prevalence of all malaria, severe malaria, and malaria-associated mortality decreased in Bangladesh after 5 years of interventions" (Haque et al. 2014, 102),

MAP 8.1

Malaria endemic districts in Bangladesh

Source: Saha et al. 2019, based on data from Bangladesh, National Malaria Elimination Program, https://nmcp.gov.bd/.
Note: There are 13 malaria endemic districts in Bangladesh: 3 high (red), 1 moderate (pink), and 9 low (blue or yellow).

FIGURE 8.5

Incidence of malaria

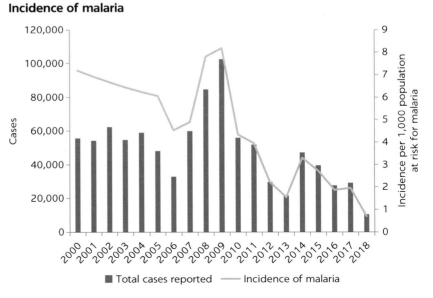

Source: Data on malaria incidence taken from the World Bank Databank, https://databank.worldbank.org.

and the decline in malaria cases across all age groups was associated with increased coverage of insecticide-treated nets. Other factors that may have contributed to this decline include effects of climate variability as average maximum temperatures in Bangladesh have risen well above the range 24°C and 28°C required for malarial parasite and vector (Haque et al. 2014). In addition, there is global evidence of malaria appearing at higher altitudes over time. This may mean that malaria prevalence will decline further, as it has reached the highest altitude in Bangladesh in the Chittagong Hill Tracts region. However, the existing risk factors for malaria include resistance to a number of drugs previously recommended for treatment for malaria as well as low socioeconomic status of large sections of the population and their close proximity to water bodies and forest areas (Islam, Bonovas, and Nikolopoulos 2013).

DIARRHEA AND CHOLERA, 1993–2018

Prevalence of diarrhea declined from 12.6 percent in 1993–94 to 4.7 percent in 2017–18 (figure 8.6), measured as the percentage of children under five years of age who had had diarrhea in the two weeks preceding the Bangladesh Demographic and Health Survey (Bangladesh NIPORT and ICF 2020).

Bangladesh has successfully managed diarrhea among children. As shown in figure 8.7, the trends in diarrhea-specific mortality and under-five child mortality caused by diarrhea are declining. In 2017–18, diarrhea accounted for only 3.25 percent of deaths among children under five years of age (Bangladesh NIPORT and ICF 2020). This is due to the significant increase in the treatment of diarrhea with oral rehydration therapy. Between 1994 and 2018, the percentage of children receiving oral rehydration therapy or increased fluids increased from 58.3 percent to 87.0 percent (Mitra et al. 1994; Bangladesh NIPORT and ICF 2020).

Cholera, a disease that has disappeared from most developed countries, continues to be endemic in 23 countries of the world, especially in tropical countries like Bangladesh with an incidence rate of 1.64 cholera cases per 1,000 population

FIGURE 8.6

Prevalence of diarrhea among children under five years of age

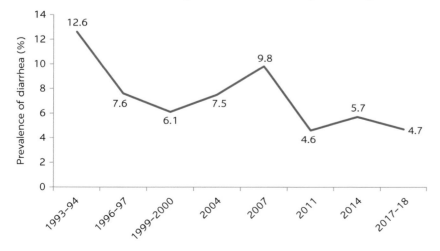

Source: Original figure for this publication.

FIGURE 8.7
Diarrhea-specific mortality and under-five mortality caused by diarrhea, 1980–2015

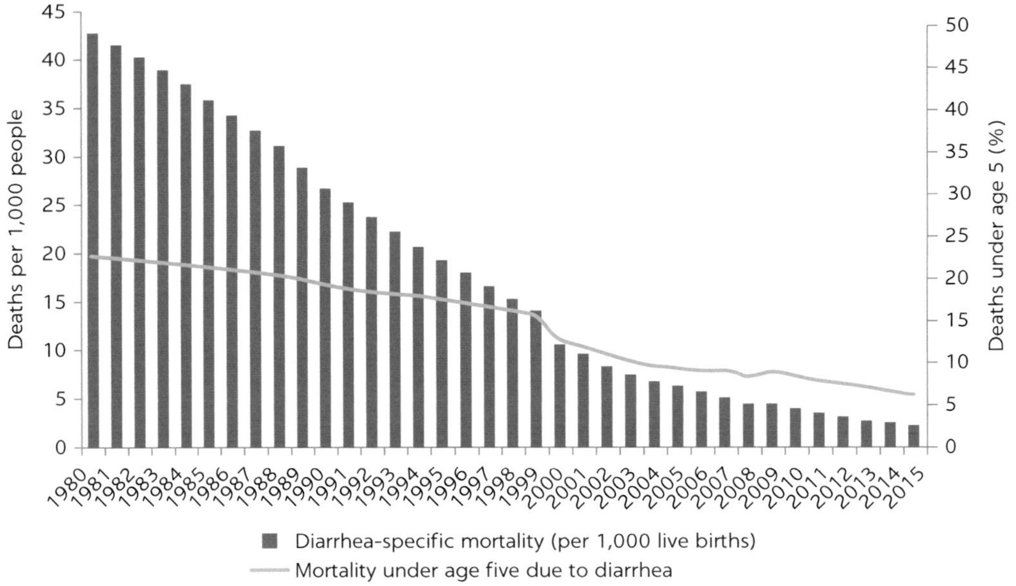

Diarrhea-specific mortality (per 1,000 live births)
Mortality under age five due to diarrhea

Source: Billah et al. 2019.

at risk (Ali et al. 2015). Within Bangladesh, cholera is endemic in certain parts of the country and its transmission increases during extreme weather events like floods and droughts. Some association of cholera transmission occurs with water temperature in ponds and rivers as well as rainfall. Although cholera affects all ages, most deaths occur in children under five years of age (Islam, Clemens, and Qadri 2018). Approximately 3,272 annual deaths are estimated to be associated with cholera in Bangladesh (Ali et al. 2015).

REFERENCES

Ali, M., A. R. Nelson, A. L. Lopez, and D. A. Sack. 2015. "Updated Global Burden of Cholera in Endemic Countries." PLOS *Neglected Tropical Diseases* 9 (6): e0003832. doi:10.1371/journal .pntd.0003832.

Bangladesh NIPORT (National Institute of Population Research and Training) and ICF. 2020. *Bangladesh Demographic and Health Survey 2017–18*. Dhaka; Rockville, MD: NIPORT and ICF. https://www.dhsprogram.com/pubs/pdf/FR344/FR344.pdf.

Billah, S. M., S. Raihana, N. B. Ali, A. Iqbal, M. M. Rahman, A. N. S. Khan, F. Karim, et al. 2019. "Bangladesh: A Success Case in Combating Childhood Diarrhea." *Journal of Global Health* 9 (2): 020803.

Haque, U., H. J. Overgaard, A. C. A. Clements, D. E. Norris, N. Islam, J. Karim, S. Roy, et al. 2014. "Malaria Burden and Control in Bangladesh and Prospects for Elimination: An Epidemiological and Economic Assessment." *Lancet Global Health 2014* 2: e98–e105.

Hsan, K., M. M. Hossain, M. S. Sarwar, A. Wilder-Smith, and D. Gozal. 2019. "Unprecedented Rise in Dengue Outbreaks in Bangladesh." *Lancet* 394 (December 14). https://www.thelancet .com/journals/laninf/article/PIIS1473-3099(19)30616-4/fulltext.

Islam, N., S. Bonovas, and G. K. Nikolopoulos. 2013. "An Epidemiological Overview of Malaria in Bangladesh." *Travel Medicine and Infectious Disease* 11 (1): 29–36.

Islam, M. T., J. D. Clemens, and F. Qadri. 2018. "Cholera Control and Prevention in Bangladesh: An Evaluation of the Situation and Solutions." *Journal of Infectious Diseases* 218 (suppl 3).

Mamun, M. A., J. M. Misti, M. D. Griffiths, D. Gozal. 2019. "The Dengue Epidemic in Bangladesh: Risk Factors and Actionable Items." *Lancet* 394 (December 14). https://www.thelancet.com /journals/lancet/article/PIIS0140-6736(19)32524-3/fulltext.

Mitra, S. N., M. N. Ali, S. Islam, A. R. Cross, and T. Saha. 1994. *Bangladesh Demographic and Health Survey, 1993–1994*. Calverton, Maryland: NIPORT, Mitra and Associates, and Macro International. https://dhsprogram.com/publications/publication-fr60-dhs-final-reports .cfm.

Mutsuddy, P., S. T. Jhora, A. K. M. Shamsuzzaman, S. M. G. Kaisar, and M. N. A. Khan. 2018. "Dengue Situation in Bangladesh: An Epidemiological Shift in Terms of Morbidity and Mortality." *Canadian Journal of Infectious Diseases and Medical Microbiology* 2019, article ID 3516284. doi:10.1155/2019/3516284.

Ooi, E. E., A. Wilder-Smith, L. C. Ng, and D. J. Gubler. 2010. "The 2007 Dengue Outbreak in Singapore." *Epidemiology and Infection* 138: 958–59, author reply 9–61. doi:10:S0950268810000026.

Saha, A., M. Sarker, M. Kabir, G. Lu, and O. Müller. 2019. "Knowledge, Attitudes, and Practices Regarding Malaria Control among the Slash and Burn Cultivators in Rangamati Hill." *Malaria Journal* 18. doi:10.1186/s12936-019-2849-0.

WHO (World Health Organization). 2016. "An Estimated 12.6 Million Deaths Each Year Are Attributable to Unhealthy Environments." Press Release March 15, 2016. Geneva: WHO. http://www.who.int/mediacentre/news/releases/2016/deaths-attributable-to -unhealthy-environments/en/.

WHO. n.d. "Information and Public Health Advice: Heat and Health." Geneva: WHO. https:// www.who.int/globalchange/publications/HeatstressAnnouncement_250818.pdf?ua=1.

WPR (World Population Review). 2021. "Population of Cities in Bangladesh (2021)." https:// worldpopulationreview.com/countries/cities/Bangladesh.

Wright, Steph. 2020. "The World's Most Densely Populated Cities." World Atlas October 4, 2020. https://www.worldatlas.com/articles/the-world-s-most-densely-populated-cities .html.

IV Way Forward and Conclusions

9 Recommendations for Public Policy

INTRODUCTION

Based on the discussions presented in parts II and III of this report, the following public policy recommendations can be implemented with the help of the existing institutions and platforms with decision makers, implementers, academia, and technical experts working in complementarity. Policy options are classified under two broad approaches: the first documents *the known* in order to be better informed with evidence and analyses on the effects of climate change and climate variability that are known to have occurred. This category includes both short- and medium-term actions and will help predict and mitigate risks. The second addresses knowing *the unknown or not-so-well-known*, an effort to explore impacts or effects that have not been adequately researched and, hence, help formulate adaptation measures over the longer term.

DOCUMENTING THE KNOWN

Record more accurate weather data with localized information

The Bangladesh Meteorological Department (BMD) needs to expand the number of weather stations geographically to be able to collect more localized and granular information on the various weather variables. Factors such as population density, migration patterns, and urbanization should play a strong role when choosing where to install additional weather stations to ensure accurate and strategic representation of these weather sites. While satellite-based information is useful to generate an overall perspective, localized climatic conditions, particularly rainfall and humidity, are important to trace climate-sensitive diseases. The effects of changing weather and consequently of climate change work in different ways: for example, the rise in temperature can cause direct and sometimes opposing effects on the ecosystems of vector-borne diseases (Fouque and Reeder 2019). Therefore, further analyses based on local climatic conditions are necessary. In addition, climate forecasts are more accurate during El Niño and La Niña episodes[1] for such areas affected by El Niño–Southern Oscillation as

Bangladesh (Lowe et al. 2017). It is, therefore, important for BMD to actively use such opportunities to record accurate data.

Strengthen surveillance of diseases and set up a "dengue early warning system"

Surveillance involves two strands of information—entomological data collection, or the study of vector or mosquito life cycles—and the information system that links epidemiological data on disease incidence with climatic data associated with transmission ecology. The Ministry of Health and Family Welfare (MoHFW) needs to strengthen its routine surveillance mechanisms to track incidence and prevalence of diseases. Second, existing literature highlights the importance of using localized climatic conditions to predict the evolution of infectious diseases, after considering other important compounding factors such as a population's immunity status, internal migration or mobility patterns, and implementation of vector control measures. In Ecuador, for example, the peak dengue season has shifted from the first trimester to the second trimester, although dengue transmission remains high in the hot and rainy season (Lowe et al. 2017). Hence, there is a need for the MoHFW to set up a climate-based dengue early warning system that will use climate data to track potential disease outbreaks. The effectiveness of such systems is dependent on the capacity to collect accurate climate information and use it to forecast patterns (Lowe et al. 2017). This points to the imperative for enhanced coordination between MoHFW and BMD to develop such a climate-based dengue early warning system.

Enhance vector control measures through innovative approaches

Aedes mosquito and dengue control in Bangladesh require a multisectoral, targeted, data-driven response. Overreliance on reactive fogging, space spraying that targets adult mosquitoes, and untargeted larval control are not an efficient use of resources. First, the government has to strengthen intra- and intersectoral collaboration. It is not just the health sector, but water, sanitation, and hygiene; education; and transportation, among others, need to be better engaged and coordinated. Effective community engagement can also help generate and disseminate correct information regarding *Aedes* control and pave the way for community-level innovations and solutions. Second, evidence suggests that mosquito control is most effective during the high temperature seasons, as the mosquito life cycle is influenced by climatic conditions (Lai 2018). Therefore, efforts for controlling mosquitoes should be strategically timed to maximize the effects of the interventions: spraying insecticides, clearing drainage, and other community-based interventions.

Measure air quality to tackle an important confounding factor

As discussed in this report, air pollution is an important determinant of respiratory illnesses and other diseases. BMD is well positioned to track air pollution levels using instruments in the field as well as satellites. Geostationary operational environmental satellites (the GOES-R series) and the joint polar satellite system monitor the particle pollution in the atmosphere. These track smoke particles from wildfires, airborne dust during dust and sandstorms, urban and industrial pollution, and ash from erupting volcanoes (SciJinks n.d.). The existing sources of information coupled with localized ground-level information can assist the government in tracking air pollution levels in Bangladesh. This information should be used to analyze how patterns of diseases are correlated and, accordingly, implement measures to reduce air pollution.

Mental health issues merit urgent attention

Due to cultural norms, mental health issues such as depression and anxiety are largely ignored in Bangladesh and merit better understanding. Once the underlying situation is better understood, effective mitigation measures should be implemented accordingly. While women are at higher risk than men for depression, men are more susceptible to anxiety. Traditional gender norms and societal underpinning, therefore, indicate the necessity of gender-sensitive programmatic efforts. In addition to a more mainstreamed response through provision of mental health services, community-based solutions for prevention and treatment are well warranted. Creation of peer support groups, for instance, can help diffuse feelings of loneliness, depression, and anxiety among the elderly. Additionally, training nonspecialists to detect and treat common mental disorders has been shown to be effective in many low- and middle-income settings. Such types of supported task-sharing approaches for mental health are worth consideration.

KNOWING THE NOT-SO-WELL-KNOWN

Further research is needed to document accurately the effects of climate change on health as well as other sectors

Establishing the causal effects of climate change on health outcomes will entail the collection and analyses of long-term, household-level data and localized climatic conditions. The Ministry of Environment, Forest, and Climate Change may be better placed to undertake such multisectoral research using household panel surveys over longer periods of time for all relevant sectors. Given their experience in dealing with large household-level surveys, the Bangladesh Bureau of Statistics may be able to assist. Such surveys would help quantify and eventually project the effects of climate change accurately, mitigate imminent risks, and identify and adapt to emerging issues. For example, it is not known if the outbreak of novel coronavirus disease in 2020 is a direct or indirect effect of climate change. Similarly, the effects of heat on human health in Bangladesh need to be understood. Also to be prioritized are more detailed research on air pollution and zoonotic diseases.

Track the carbon footprint for Bangladesh and limit greenhouse gas emissions

While the country's vulnerability to the effects of climate change is well known—Bangladesh bears a disproportionate brunt of the world's increased greenhouse gas emissions—what is not known is Bangladesh's contribution to the problem of climate change. In 2012, Bangladesh emitted 190 million metric tons of greenhouse gas, which represents almost a 60 percent increase between 1990 and

2012, and the agriculture industry and energy sector contributed to over 70 percent of the total emissions in 2012 (ClimateLinks 2016). The Government of Bangladesh's Climate Change Action Plan identifies measures to reduce carbon emissions. It is now critical to ensure that such measures are implemented to help limit the country's total carbon emissions.

NOTE

1. "El Niño is the warm phase of the El Niño–Southern Oscillation (ENSO) and is associated with a band of warm ocean water that develops in the central and east-central equatorial Pacific. The ENSO is the cycle of warm and cold sea surface temperature of the tropical central and eastern Pacific Ocean. El Niño is accompanied by high air pressure in the western Pacific and low air pressure in the eastern Pacific. El Niño phases are known to last close to four years, but can continue for any duration between two and seven years. The cool phase of ENSO is La Niña, with sea surface temperature in the eastern Pacific below average, and air pressure high in the eastern Pacific and low in the western Pacific. The ENSO cycle, including both El Niño and La Niña, causes global changes in temperature and rainfall." As defined in *Wikipedia*, "El Niño," https://en.wikipedia.org/wiki/El_Ni%C3%B1o.

REFERENCES

ClimateLinks. 2016. "Greenhouse Gas Emissions Factsheet: Bangladesh." Washington, DC: US Agency for International Development. https://www.climatelinks.org/resources/greenhouse-gas-emissions-factsheet-bangladesh.

Fouque, F., and J. C. Reeder. 2019. "Impact of Past and On-going Changes on Climate and Weather on Vector-borne Diseases Transmission: A Look at the Evidence." *Infectious Diseases of Poverty* 8 (51). doi:10.1186/s40249-019-0565-1.

Lai, Y-H. 2018. "The Climatic Factors Affecting Dengue Fever Outbreaks in Southern Taiwan: An Application of Symbolic Data Analysis." *Biomedical Engineering Online* 17 (suppl 2): 148. doi:10.1186/s12938-018-0575-4.

Lowe, R., A. M. Stewart-Ibarra, D. Petrova, M. Garcia-Díez, M. J. Borbor-Cordova, R. Mejía, M. Regato, and X. Radó. 2017. "Climate Services for Health: Predicting the Evolution of the 2016 Dengue Season in Machala, Ecuador." *Lancet Planet Health 2017* 1: e142-e151.

SciJinks. n.d. "How Is Air Quality Measured?" SciJinks. https://scijinks.gov/air-quality/#:~:text=Air%20quality%20is%20measured%20with,of%20pollution%20in%20the%20air.

10 Conclusions

CLIMATE VARIABILITY AND DISEASES IN BANGLADESH

Bangladesh's vulnerability to the deteriorating effects of climate change is well documented. Because of its geographical attributes and location, Bangladesh is listed among the 10 countries in the world most vulnerable to climate change. Evolving climatic conditions have already negatively affected human health, and the effects are expected to intensify with the predicted changes in weather patterns. For instance, in tropical countries such as Bangladesh, infectious disease transmission is likely to precipitate certain vector-borne diseases such as malaria or dengue fever and waterborne diseases such as diarrhea and cholera. Incidences of respiratory diseases can increase due to extreme temperatures that aggravate airborne allergens and pollution (World Bank 2012). A global rise in temperature by 4°C—a scenario referred to as the worst case of global warming[1]— would create unprecedented levels of stress on health that would likely overburden the global health systems to a point where adaptation would no longer be feasible (World Bank 2012). Hence, the prevailing situation underscores the urgency for the public sector to be better prepared to respond to the crisis.

TRACKING AND UNDERSTANDING CORRELATIONS OF WEATHER PATTERNS TO HUMAN HEALTH IN BANGLADESH

There is a need to better understand how the climate has changed over the years and to document the connections between climate variability and human health. Climate variability refers to short-term changes in weather variables on average; these may occur over a month, a season, or a year. Climate change, however, refers to changes in average meteorological conditions and seasonal patterns over a longer time, often covering several years (Mani and Wang 2014). Globally, literature on quantifying the effects of climate variability on health is limited. Evidence from Bangladesh is constrained by analyses that focus on specific regions in the country; are conducted with small nonrepresentative datasets, health conditions, or both; and are generally not related to exposure to real-time climatic data.

This report, therefore, has sought to fill the knowledge gap across four broad areas: (1) use documented evidence to establish connections between climate change or climate variability and health, as well as the influence of weather variables, particularly on mosquitoes, one of the largest sources of vector-borne diseases; (2) study changes in weather patterns in Bangladesh historically since 1901 with a focus on the past 44 years; (3) gather primary data representative of urban and rural areas to quantify the relationship between climate variability and infectious diseases and mental health in Bangladesh;[2] and (4) apply globally recognized standards to measure the prevalence of depression and anxiety at the urban and rural levels and establish its causal relationship to climate variability and seasonal patterns in Bangladesh.[3]

CLIMATE VARIABILITY'S IMPACTS ON THE INCIDENCE OF INFECTIOUS DISEASES AND MENTAL HEALTH

The report provides two pieces of evidence. First, it finds that humidity and mean temperature are negatively correlated to waterborne diseases but positively correlated to respiratory illnesses from analyses of primary data collected from a representative sample of 3,600 households. Likewise, mean humidity is positively correlated to mental health issues of anxiety and depression, while mean temperature is negatively associated with depression.

Second, data related to the 2019 dengue outbreak in Dhaka indicate that climate variability played a key role. The outbreak of dengue cases can be partially explained by weather patterns—uncharacteristically heavy rain in February 2019, followed by temperature and humidity levels conducive to mosquito breeding in the subsequent months from March to July 2019.

A TIPPING POINT IN BANGLADESH

Considerable changes to the climate have already occurred in Bangladesh. Over the past 44 years, Bangladesh has become hotter, with a 0.5°C increase in mean temperature recorded between 1976 and 2019. Trend analyses indicate that the maximum temperature continues to rise for all months except December and has already substantially increased from February to November. Overall, summers are becoming hotter and longer with the monsoon period being extended from February and October, while winters are becoming warmer. Bangladesh appears to be losing its distinct seasonality.

IMPLICATIONS AND NEXT STEPS FOR BANGLADESH

The projected changes in climate will have considerable ramifications on the health of the population. With further climatic changes predicted across Bangladesh, including increases of temperature by approximately 1.4°C circa 2050 and annual rainfall expected to rise by 74 millimeters by 2040–59, the deleterious effects on human physical and mental health are likely to escalate. These projections and the overall findings of the report underscore the need for (1) improving data collection systems for better predictability of weather and associated impacts on health outcomes, (2) strengthening health systems to preempt and mitigate potential outbreaks of infectious and other emerging or reemerging climate-sensitive diseases, and (3) ensuring adequacy of response mechanisms for better adaptation to the effects of climate change.

NOTES

1. Presented at the Paris Climate Change Conference of Parties in 2015 where the Paris Agreement was signed.
2. Sampling also allows for disaggregated analysis for major urban centers such as Dhaka and Chattogram cities.
3. Using the locally contextualized versions of the tools, depression is measured using the Patient Health Questionnaire-9 (PHQ-9) and anxiety using Generalized Anxiety Disorder-7 (GAD-7).

REFERENCES

Mani, M., and L. Wang. 2014. "Climate Change and Health Impacts: How Vulnerable Is Bangladesh and What Needs to Be Done." *End Poverty in South Asia* (blog), May 7, 2014. https://blogs.worldbank.org/endpovertyinsouthasia/when-climate-becomes-health-issue -how-vulnerable-bangladesh.

World Bank. 2012. *Turn Down the Heat: Why a 4°C Warmer World Must Be Avoided.* Washington, DC: World Bank. https://openknowledge.worldbank.org/handle/10986/11860.

Detailed Review of Existing Literature Linking Climate Change to Diseases, Infections, and Illnesses Globally and in Bangladesh

INFECTIOUS DISEASES: GLOBAL PERSPECTIVE

Climate change will affect many climate-sensitive infectious diseases through the survival, reproduction, or distribution of disease pathogens and hosts as well as the availability and means of their transmission environment. Human behavior such as crowding and displacement will amplify risks of infection (McMichael 2012). An agent (pathogen), a vector (host), and favorable transmission environment are three components that are essential for infectious diseases to spread (Wu et al. 2016). There is a limited range of climatic conditions—the climate envelope—within which each infective agent or vector species can survive and reproduce (Patz et al. 2003). Therefore, changes in the climate, which include alterations in one or more variables like temperature, rainfall, sea-level elevation, wind, and duration of sunlight, directly impact the survival, reproduction, or distribution of disease pathogens and hosts, as well as the availability and means of their transmission environment (Wu et al. 2016).

Sufficient observational evidence is available on the effects of meteorological factors on the incidence of vector-borne, waterborne, airborne, and foodborne diseases. A large number of studies have been conducted globally to identify the seasonal patterns and climatic sensitivities of many of these infectious diseases. A more contemporary concern is the extent to which changes in disease patterns will occur under the conditions of global climate change (Patz et al. 2003). As such, the correlation between meteorological factors and the components of transmission cycles—for example, parasite development rates, vector biting and survival rates—or the observed geographical distribution of disease have been used to generate predictive models (Campbell-Lendrum et al. 2015). These models link projections of future scenarios of climate change with other determinants like gross domestic product—as a measure of socioeconomic and technological development—and urbanization. However, because of uncertainties in climate projections and future development trends as well as the compounding effects of natural climate variability over short to medium timescales—from years to one or two decades—the models are highly approximate and are only able to comment on broad trends, for example, the effects of climate changes in populations at risk at the global scale for the 2030s or the 2050s, rather than at local levels for specific years (Campbell-Lendrum et al. 2015).

Malaria

Globally, malaria, dengue, yellow fever, chikungunya, and Zika are the major vector-borne diseases that are a concern of public health, as these diseases are associated with a high burden of morbidity and mortality. These are now mostly transmitted human to human via *Anopheles* and *Aedes* mosquitoes in tropical and subtropical regions (Ogden 2017). *Aedes* can spread viruses that cause dengue fever, yellow fever, and Zika, while *Anopheles* can spread parasites leading to malaria. Tropical and subtropical climatic regions of the world including Africa, Asia, Central and South America, and certain Caribbean islands are highly conducive to malaria transmission given the temperature and humidity needs of the *Anopheles* mosquitoes and *Plasmodium* parasites (Sadoine et al. 2018; Nabi and Qader 2009).

The relationship between malaria transmission, climatic factors, and socioeconomic conditions is complex. The number of cases of malaria has dramatically decreased over the past decades due to large investments in curbing malaria. However, given current trends, climatic conditions in the future may become more suitable for malaria transmission in the tropical highland regions (Flahault, de Castaneda; and Bolon 2016; Caminade et al. 2014). Studies combining data from different climatic zones with simulations suggest that climate change will increase malaria burden (Campbell-Lendrum et al. 2015; Hay et al. 2006; Caminade et al. 2014). Caminade et al. (2014) evaluated three malaria outcome metrics at global and regional levels: climate suitability, additional population at risk, and additional person-months at risk. The malaria projections were based on five different global climate models, each run under four emission scenarios—representative concentration pathways (RCPs)—and a single population projection. The findings show an overall global net increase in climate suitability and a net increase in the population at risk, when comparing from RCP 2.6 to RCP 8.5 from the 2050s to the 2080s, but with large uncertainties (Caminade et al. 2014). However, if poverty outcomes continue to improve, investments in the health sector are maintained, disease-specific interventions are maintained, and risks associated with insecticide and drug resistance are well managed, it should be possible to drive down global malaria rates in the future (Campbell-Lendrum et al. 2015). For example, in Africa, an estimated 663 million clinical cases of malaria were averted between 2000 and 2015 through advancement in malaria control and prevention through insecticide-treated bed nets, artemisinin-based combination therapy, and insecticide residual spraying (Sadoine et al. 2018). Nevertheless, climate change may continue having effects in specific geographical areas where the protective factors above are not maintained (Campbell-Lendrum et al. 2015). Increased insecticide resistance, land use changes, population mobility, and population growth with inadequate housing are associated with rising incidence and are likely to modify the relationship between climate variability and malaria (Sadoine et al. 2018).

Dengue

The global incidence of dengue has grown dramatically over the past several decades. With an estimated 390 million infections each year, nearly half of the world's population is at risk (WHO 2019). Dengue is typically found in tropical and subtropical climates, mostly in urban and periurban areas. The endemic regions include Southeast Asia, Latin America, Asia, and the Caribbean (Raheel et al. 2010). Climatic factors such as rainfall, temperature, and humidity play a pivotal role in the epidemiological pattern of dengue in the number of cases, severity of illness, shifts in affected age groups, and the expansion of spread from urban to rural areas (Raheel et al. 2010; Sirisena and Noordeen, 2013).

Studies modeling the relationship between climate change and dengue show that trends in climate change have continued and will continue to provide favorable conditions for the disease's transmission (Campbell-Lendrum et al. 2015). Evidence highlighting the preventive effects of either socioeconomic development or specific disease control measures is much weaker for dengue than for malaria. As such, incidence of dengue is high in many regions where malaria has been effectively controlled or eradicated (Ooi et al. 2010). The incidence is also higher in urban areas, especially in slums typically characterized by poor and dense housing and limited sanitation and waste management facilities. In addition, nonclimatic factors like rapid urbanization and increases in international trade and travel will likely intensify the effects of climate change (Campbell-Lendrum et al. 2015).

Respiratory illnesses

Climate alterations may directly or indirectly affect the incidence and severity of respiratory infections by affecting the disease vectors and the host immune responses (Mirsaeidi et al. 2015). The effects range from direct heat and air pollution to changes in the biological burden of allergens and shifting infectious disease patterns (Takaro, Knowlton, and Balmes 2013). Many epidemiological studies have established relationships between rising atmospheric temperatures associated with climate change and evolving distributions of respiratory infections and mortality (Takaro, Knowlton, and Balmes 2013; Budiyono, Jati, and Ginandjar 2017; Mirsaeidi et al. 2015). Increasing atmospheric temperatures may exacerbate ground-level pollution, notably ozone, which affects lung function and increases acute premature mortality, asthma-related hospitalizations, and emergency department visits (Takaro, Knowlton, and Balmes 2013). Chronic ozone exposures also increase mortality risks in people with existing chronic obstructive pulmonary disease, cardiovascular diseases, and diabetes (Takaro, Knowlton, and Balmes 2013; Mirsaeidi et al. 2015). Simulations show that by the year 2050, as many as 2,500 summertime deaths may be attributable to premature ozone-related mortality associated with climate change under a high-emissions scenario, absent other limits on ozone precursors (Takaro, Knowlton, and Balmes 2013).

The variations in the incidence of respiratory infections are seasonal. Lower respiratory tract infections have a higher incidence during the winter in temperate areas. For example, both influenza and streptococcal pneumonia have higher incidence during winter months in the United States and are among the top 10 causes of death (Mirsaeidi et al. 2015). In the tropical and subtropical areas of Asia and Africa, incidences of pneumonia are usually higher during the monsoon, demonstrating the association of temperature, humidity, and rainfall with pneumonia patterns (Takaro, Knowlton, and Balmes 2013; Mirsaeidi et al. 2015). While the exact cause of pneumonia seasonality is unclear, indoor crowding, lower relative humidity, seasonal variation in the human immune system, association with other common seasonal respiratory pathogens, indoor air pollution, and low ultraviolet radiation might all contribute to pneumococcal disease risk (Mirsaeidi et al. 2015).

INFECTIOUS DISEASES: THE BANGLADESH CONTEXT

In Bangladesh, temperature and precipitation changes have been found to impact the dynamics of vector-borne diseases such as malaria, dengue, visceral leishmaniasis—commonly known as kala-azar—cholera, and diarrheal diseases (Rahman et al. 2019; Banu et al. 2014; Hossain, Noiri, and Moji 2011; Reid et al. 2012; Hashizume et al. 2007). Although the country has made progress in controlling communicable diseases in recent years, an upsurge in dengue cases and more recently an increase in chikungunya and Zika cases, are causing major threats to the health of the population. Higher temperatures are expected to increase the transmission and spread of vector-borne diseases by increasing mosquito density in some areas and increasing replication rate and bite frequency of mosquitoes (Costello et al. 2009). This will, in turn, increase the incidence of malaria, dengue, and tick-borne encephalitis. In Bangladesh, very few studies have explored the relationship between environmental variables and infectious diseases. Chowdhury et al. (2018) investigated six climate-sensitive infectious diseases—malaria, diarrheal disease, enteric fever, encephalitis, pneumonia, and bacterial meningitis—using hospital-based data over five years to analyze the possible relationships with common climatic variables. All six diseases, particularly pneumonia, showed highly significant rises in incidence between 2008 (540 cases) and 2012 (1,330 cases) compared to overall all-cause hospital admissions over the same period. The highest number of malaria, diarrhea, and pneumonia cases occurred during the rainy season. High cases of enteric fever, encephalitis, and meningitis cases occurred during autumn, which follows the rainy season. Increased temperature was positively associated with the incidence of malaria, enteric fever, and diarrhea. Malaria and diarrhea were also positively associated with higher humidity.

Dengue and chikungunya

Driven by its high burden of morbidity and mortality, dengue is emerging as a serious public health problem in Bangladesh (Mone 2019). The infection is now reaching all parts of the country, particularly affecting children, pregnant women, and older people (Mone 2019). Poor sanitation and hygiene systems, along with high population density and lack of proper mosquito-control strategies, contribute to the high disease burden. Table A.1 summarizes some of the key studies assessing the relationship between climatic variables and dengue in Bangladesh.

TABLE A.1 **Key studies that assess the relationship between climatic variables and dengue in Bangladesh**

SOURCE	POPULATION AND DATA	MAIN FINDINGS
Sharmin et al. 2015. "Interaction of Mean Temperature and Daily Fluctuation Influences Dengue Incidence in Dhaka, Bangladesh"	Monthly dengue cases in Dhaka district, between January 2000 and December 2009, were obtained from the Directorate General of Health Services (DGHS). Population estimates were extracted from the 1991, 2001, and 2011 census data of the Bangladesh Bureau of Statistics.	Significant associations were found between mean temperature and temperature fluctuation and dengue incidence, adjusted for rainfall and population density over a period of 10 years in Dhaka. High mean temperature with low fluctuations increased dengue incidence one month later. Besides temperature, sea surface temperature anomalies and population density exerted a significant positive influence on dengue incidence, indicating increasing risk of dengue in the overpopulated city.
Mutsuddy et al. 2018. "Dengue Situation in Bangladesh: An Epidemiological Shift in Terms of Morbidity and Mortality"	40,476 dengue cases in Bangladesh, between 2000 and 2017, were obtained from the DGHS.	Average rainfall, humidity, temperature, and rapid unplanned urbanization were strong predictors of an increase in dengue cases in 2016 and the emergence of the chikungunya virus for the first time in Bangladesh in 2017. Despite efforts to control dengue based primarily on vector control and case management, the burden and costs of the disease and similar vector-borne diseases are likely to continue to grow in the future.
Banu et al. 2014. "Projecting the Impact of Climate Change on Dengue Transmission in Dhaka, Bangladesh."	25,059 dengue cases in Dhaka city, between January 2000 and December 2010, were obtained from the DGHS	The study showed that weather variables—particularly temperature and humidity—were positively associated with dengue transmission. The effects of weather variables were observed at a lag of four months. Assuming a temperature increase of 3.3°C without any adaptation measures and changes in socioeconomic condition, there will be a projected increase of 16,030 dengue cases in Dhaka alone by the end of this century.
Hashizume et al. 2012. "Hydroclimatological Variability and Dengue Transmission in Dhaka, Bangladesh: A Time-Series Study"	3,130 dengue cases in Dhaka city, between January 2005 and December 2009, were obtained from 11 major hospitals.	The effects of river levels and rainfall on hospital admissions for dengue fever were analyzed at 11 major hospitals in Dhaka. There was strong evidence for an increase in dengue fever at high river levels. Hospitalizations increased by 6.9% (95% CI: 3.2, 10.7) for each 0.1-meter increase above a threshold (3.9 meters) for the average river level over lags of 0–5 weeks. Conversely, the number of hospitalizations increased by 29.6% (95% CI: 19.8, 40.2) for a 0.1-meter decrease below the same threshold of the average river level over lags of 0–19 weeks. The findings provide evidence that factors associated with both high and low river levels increase the hospitalizations of dengue fever cases in Dhaka.
Ahmed et al. 2007. "Seasonal Prevalence of Dengue Vector Mosquitoes in Dhaka City, Bangladesh"	*Aedes* larvae were collected from all the 90 wards of Dhaka city, between December 2001 and August 2002.	The seasonal prevalence of *Aedes aegypti* and *A. albopictus* was analyzed in Dhaka city, from December 2001 to August 2002. *A. aegypti* and *A. albopictus* were active in both dry and wet seasons with a peak in July, when the rainfall was the highest. From September to April, the larval population level remained low. Reduction of larval population during winter months was related to low rainfall.
Islam et al. 2018. "Correlates of Climate Variability and Dengue Fever in Two Metropolitan Cities in Bangladesh"	22,970 dengue cases (21,748 in Dhaka city and 1,222 in Chattogram city) between January 1, 2000, and December 31, 2009, were obtained from the DGHS.	The correlation between temperature, humidity, and rainfall with dengue fever were analyzed in two dengue endemic cities—Dhaka and Chattogram—through time series analysis from January 2000 to December 2009. Mean dengue incidence was 31.62 (standard deviation 28.7) per 100,000 population in Dhaka whereas it was 5.76 (standard deviation 11.7) per 100,000 in Chattogram. The incidence of dengue cases was found significantly associated with monthly mean temperature, total rainfall, and mean humidity in Dhaka, although in Chattogram, the significantly associated factors were monthly total rainfall and mean humidity.

Source: Original table for this publication.
Note: Full source citations are found in the chapter reference list. CI = confidence interval.

Four studies have reported a positive association between mean temperature and dengue (Banu et al. 2014; Sharmin et al. 2015; Islam et al. 2018; Mutsuddy et al. 2018). Islam et al. (2018) report using a time-series analysis that showed that the incidence of dengue cases was significantly associated with monthly mean temperature, total rainfall, and mean humidity in Dhaka. In Chattogram, however, monthly total rainfall and mean humidity played a bigger role. Ahmed et al. (2007) found that although *A. aegypti* and *A. albopictus* were active in both dry and wet seasons, they peaked in July, during the time with the highest rainfall. From September to April, the larval population level remained low, and the reduction of larval population during winter months was related to low rainfall (Ahmed et al. 2007). Another study reported on the association between river levels and rainfall on incidence of dengue (Hashizume et al. 2012). The authors suggest that rates of hospitalizations increased by 6.9 percent for each 0.1-meter increase in river levels due to rainfall above a critical threshold (3.9 meters) for the average river level over lags of 0–5 weeks.

Sharmin et al. (2015) conducted a study over a 10-year period in Dhaka and found that high population density positively influenced dengue incidence, indicating increasing risk of dengue in the overpopulated city. Lack of regular water supply, precarious reservoirs for potable water, and public garbage collection practices create potential breeding sites for mosquitos, which may help explain the high incidence of dengue affecting thousands each year (Tauil 2001). Severe dengue, if not appropriately managed, may lead to rapid death, particularly in children (Phakhounthong et al. 2018). Maternal dengue infection has been linked to several adverse health effects and pregnancy outcomes including vertical transmission to the fetus, preterm birth, low birth weight, preeclampsia and eclampsia, Caesarean delivery, and fetal or perinatal death and maternal death (Rylander, Odland, and Sandanger 2013).

Mutsuddy et al. (2018) discuss the economic burden of dengue. The disease's sizable prevalence and low public spending on health in general coupled with a rising burden of out-of-pocket spending—up to 67 percent, which is the highest in Southeast Asia—are creating an increasingly insurmountable burden on the populace. Based primarily on vector control and case management, efforts are being put into controlling dengue, but the burden and costs of the disease and similar vector-borne diseases are expected to continue to grow in the future (Mutsuddy et al. 2018).

Recent emergence of chikungunya viruses, spread by the *Aedes* mosquitoes, has been an added source of concern for the country. Average rainfall, humidity, temperature, and rapid unplanned urbanization were found to be strong predictors of dengue cases in 2016 and the emergence of the chikungunya virus for the first time in Bangladesh in 2017. The sudden outbreak of chikungunya was mainly attributable to unusually excessive rainfalls from January to March 2017, creating suitable climatic conditions for the *Aedes* vector (Kabir et al. 2016). From the onset of the outbreak between April and September 2017, the Ministry of Health and Family Welfare reported more than 13,176 clinically confirmed cases in 17 of 64 districts, with Dhaka particularly affected. Another study combined mobility data with epidemiological data from a household survey to understand the impact of large-scale population movements on the spatial spread of the chikungunya virus within and outside the city (Mahmud et al. 2019). They found that the peak of the 2017 chikungunya outbreak in Dhaka coincided with the annual Eid holidays, during which large numbers of people traveled from Dhaka to their native region in other parts of the country.

Malaria

A number of studies have shown that meteorological factors affect adult mosquito abundance by altering the quality and quantity of breeding habitats (Bashar and Tuno 2014). Three weather parameters—temperature, humidity, and rainfall—are critical for mosquito activity and malaria epidemiology (Amin, Tareq, and Rahman 2011). A study analyzing data from 1972 to 2002 in Rangamati, Sylhet, and Faridpur found that a rise of annual average maximum temperature, yearly total rainfall, and annual average humidity are positively associated with malaria prevalence (Amin, Tareq, and Rahman 2011). Bashar and Tuno (2014) found that a lag in rainfall and relative humidity were the most significant variables influencing four species of anopheles, which are positively related to malaria cases. The same study also reported that the effects of temperature did not significantly affect the abundance of *Anopheles* mosquitoes in Bangladesh. Another study done in Chattogram Hill Tracts found that women are more vulnerable to malaria, and pregnancy is a risk factor for asymptomatic *Plasmodium falciparum* infection development (Khan et al. 2014). Chowdhury et al. (2018) conducted a tertiary-facility-based observational study to link the effect of climate change on infectious diseases in admitted-patient data in northeast Bangladesh between 2008 and 2012. They found a positive correlation between increased temperature and the incidence of malaria; higher humidity correlated with a higher number of cases of malaria; and increase in malaria incidences with increased rainfall.

WATERBORNE DISEASES

Bangladesh suffered extensive flooding during the monsoons of 1988, 1998, and 2004. During these periods, 25–50 percent of Bangladesh was submerged, resulting in the destruction of infrastructure, water contamination, and an epidemic of diarrheal illnesses leading to a substantial number of deaths (Schwartz et al. 2006). After floods, outbreaks of waterborne diseases such as diarrhea were thought to result primarily from contamination of water caused by disruption of water purification and sewage disposal systems. Secondary effects of flooding including crowding and subsequent fecal-oral spread of gastrointestinal pathogens contribute to the spread of diarrheal diseases. During the 1988 flood in Bangladesh, diarrheal disease was responsible for 35 percent of all flood-related illnesses and about 27 percent of all flood-related deaths in rural Bangladesh (Siddique et al. 1989). During the 1998 flood, 25 percent of all people surveyed in two rural areas of Bangladesh reported diarrheal illnesses associated with floods (Kunii et al. 2002).

Cholera

Cholera transmission increases during both floods and droughts, and water temperature in ponds and rivers, in addition to rainfall, have a remote association with cholera transmission (Islam et al. 2018). Climate change is seen to have the potential to increase the prevalence and magnitude of cholera outbreaks. One study examined the relationship between cholera and heat waves in Matlab between January 1983 and April 2009 (Wu et al. 2018). It found a higher risk of cholera two days after heat waves during wet days. For households with less medium-dense tree cover, the heat wave after a two-day lag was positively associated with the risk of cholera during wet days. These findings suggest that heat waves might promote the occurrence of cholera, while this relationship was modified by rainfall and tree cover.

ZOONOTIC DISEASES

Increased urbanization and changing climate have led to the rise in epidemics of new zoonotic diseases affecting human health every year (Rana and Singh 2015). For example, outbreaks of leptospirosis, a zoonotic disease spread via skin contact with contaminated water and soil, usually through the urine of domestic or wild animals, can occur during floods, especially in overcrowded areas. In Bangladesh, the risk of leptospirosis increases during the rainy season (Patwary, Bari, and Islam 2016). Extreme weather events such as cyclones and floods are expected to occur with increasing frequency and greater intensity and may potentially result in an upsurge in the disease incidence as well as the magnitude of leptospirosis outbreaks in the future (Lau et al. 2010).

Bangladesh experienced the first outbreak of Nipah virus encephalitis, a zoonotic disease that causes potentially lethal inflammation of the brain, in 2001. Since then there have been approximately 33 recorded outbreaks over the subsequent 13 years. Nipah virus is typically transmitted by bats, but it is also susceptible to peer-to-peer transmission. In Bangladesh, Nipah virus is usually acquired through consumption of date palm sap contaminated with bat saliva or urine. Many ecological factors contribute to the emergence of Nipah virus. However, the most prominent is human intervention into the bat infested areas due to rapid urbanization (Rana and Singh 2015). It has been predicted that climate change may shift the current distribution of flying foxes (a breed of bats), introducing the virus to previously unexposed areas (Rahman et al. 2019). Extreme weather events, such as heat waves, could increase the risk of transmission to humans by placing bat populations under physiological stress that could trigger prolonged viral shedding (Rahman et al. 2019).

MENTAL HEALTH: GLOBAL PERSPECTIVE

The American Psychological Association Task Force concluded that climate-change-related psychological health impacts will likely be widespread, cumulative, and profound (Swim et al. 2009). The impacts will be experienced most acutely by those with preexisting mental illnesses, marginalized populations, communities who rely on the local ecosystems, and those living in areas most susceptible to climate change (Berry 2009). McMichael, Woodruff, and Hales (2006) refer to the mental health impacts as a deferred risk of climate change and include the psychological impacts associated with rural to urban displacement and the mental health consequences of droughts in failing rural communities.

According to Berry, Bowen, and Kjellstrom (2010), climate change may affect mental health directly by exposing people to psychological trauma associated with higher frequency, intensity, and duration of climate-related disasters, including extreme heat exposure. The destruction of property and landscape diminishes people's sense of belonging and their connectedness to the land. The indirect effects are via physical health—for example, extreme heat exposure, injuries, and diseases—and community well-being (Berry, Bowen, and Kjellstrom 2010). Vulnerable people and places, especially in low-income countries, are particularly affected. Adverse mental health outcomes in the aftermath of natural disasters include, among others, posttraumatic stress disorder, major depression, and somatoform disorders (Page and Howard 2010). While enhancing disaster preparedness has become an international priority in recent years, the psychological implications of disasters are often underrecognized and there is limited research on this issue (Costello et al. 2009).

Obradovich et al. (2018) found that short-term exposure to extreme weather, multiyear warming, and tropical cyclone exposure are each associated with worsened mental health. The study coupled meteorological and climatic data with reported mental health challenges drawn from nearly 2 million randomly sampled US residents between 2002 and 2012, and found that shifting from monthly temperatures between 25°C and 30°C to more than 30°C increases the probability of mental health challenges by 0.5 percentage points, that 1°C of five-year warming is associated with a 2 percentage point increase in the prevalence of mental health issues, and that exposure to Hurricane Katrina is associated with a 4 percentage point increase in this metric (Obradovich et al. 2018).

MENTAL HEALTH: THE BANGLADESH CONTEXT

Natural disasters and environmental degradation caused by climate change or climate variability are known risk factors that can affect the psychological health of vulnerable populations in Bangladesh, especially those living in coastal areas— although this has not been documented well in the local context. The majority of the people living in Bangladesh's coastal areas are low-income agricultural workers, many of whom are landless and relatively asset poor (GOB 2008; Paul 2009). They are frequently affected by natural disasters but have insufficient resources to protect themselves and adequately rebuild their lives after the event or to access the medical services when needed (Nahar et al. 2014). The initial response is to ensure that survivors receive the basic necessities to sustain life, such as shelter, food, safe water, and sanitation. However, after this acute, emergency phase, many of the affected populations or climate refugees are left with some level of psychological or mental health problems (Nahar et al. 2014). Some of these include posttraumatic stress disorder (PTSD), depressive symptoms or major depressive disorders, anxiety or generalized anxiety disorders, as well as more general mental health problems such as sleep disruption, substance abuse, and aggression (Norris 2006; Paul 2009).

A survey conducted after Cyclone Sidr in 2007, in which 10 million people were affected and 1.5 million houses were damaged, found that 25 percent of the survivors suffered from PTSD, 18 percent had major depressive disorders, 16 percent suffered from somatoform disorders, and 15 percent had mixed anxiety and depressive disorders (WEDO 2008; Paul 2009). A study conducted in the Tangail district of Bangladesh found that four months after a tornado had killed 600 people, 66 percent of the disaster-affected people were traumatized and in need of psychological support (Choudhury, Quraishi, and Haque 2006). A study exploring the damaging effects of climate change on psychological health of people living in the Hill-Tracts region of Bangladesh found that impacts of climate change can be felt both at the individual and community levels, with psychological health out-comes ranging from psychological distress, depression, and anxiety, to increased addictions and suicide rates (Kabir 2018). In Bangladesh, women are more vulner-able than men at every stage of a disaster (WHO 2002). This can be largely attributed to limited access to critical services and facilities during and after the disaster, household responsibilities, sexual harassment, the consequences of wid-owhood, and socioeconomic status (Nahar et al. 2014; WEDO 2008).

VARIABILITY AND MOSQUITO LIFE CYCLE

The patterns of climate change observed for Bangladesh over the last 44 years can be linked to the climate suitability for mosquitoes, keeping in view the correlation between weather variables and incidence of dengue, as described in chapter 2.

Costa et al. (2010) find important links between temperature and humidity and the mosquito life cycle: the mosquito population increases in warm, rainy seasons while decreasing in drier periods. Egg production of mosquitoes is high at tem-peratures of 25°C and humidity at 80 percent, compared to 35°C and 60 percent humidity when the number of eggs produced falls significantly; this suggests that the population of mosquitoes is reduced with high temperatures and low humid-ity. Second, they conclude that at mild temperatures of 25°C, there is a 43 percent increase in the number of eggs as the period of oviposition gets extended, com-pared to high temperatures of 35°C; this indicates that the mosquito population can double at mild temperatures in hot climatic conditions. Third, they found that the life span of female mosquitoes was extended up to 11 days at 25°C and 80 per-cent relative humidity but was reduced to half at 35°C regardless of humidity; this implies that at higher temperatures the female mosquitoes survive less at higher temperatures and lower humidity because of dehydration. Fourth, they found that the hatching rate of mosquito larvae from eggs increases with increases in tem-perature from 25°C to 30°C at 80 percent relative humidity, and a lower number hatched with rising temperatures at 60 percent humidity; this indicates high tem-peratures couple with low humidity cause a decrease in hatching rate.

Fouque and Reeder (2019) conclude that suitable temperature conditions are primary factors that determine disease transmission—even if competent species and vectors are present, the pathogen will not spread if temperatures are not suitable. They illustrate that although numerous cases of dengue, chikungunya, and Zika virus are imported in European countries from travelers, these do not result in frequent local transmission even though the vector or mosquitoes are present, because the temperatures are not in the favorable range of 20°C to 35°C. Temperature affects the extrinsic incubation period—which is the period of the pathogen amplifying or circulating within the insect's body—as well as the biting behavior of mosquitoes. With rising temperatures, the extrinsic incubation period shortens (Colón-González, Lake, and Hunter 2013). An adult mosquito gets infected by a pathogen through a blood meal from an infected person; the pathogen then amplifies within the insect's body before the mosquito becomes infected; and thereafter it can bite other humans to spread the disease (Fouque and Reeder 2019). Although dengue cases increase with human population growth, Fouque and Reeder (2019) conclude that incidence of dengue can increase with an increase in temperature even without population growth.

Lowe et al. (2017) found that transmission of dengue and other types of arbo-viruses spread by *Aedes* mosquitoes occur between 18°C and 34°C, with maximal transmission in the range of 26°C to 29°C. They also found that rainfall and drought can increase the availability of mosquito larvae habitats, which is containers with standing water, depending on the type of water storage practices and infrastructure for piped water. More recent research by Karkarla et al. (2019) based on data from India indicate that dengue is most significantly impacted by minimum temperature of 26°C, maximum temperature of 32°C, with 0–5 weeks' lag, and rainfall of 60 millimeters with a lag of 8–12 weeks. Their research shows with an increase of cumulative weekly rainfall from 40 to 60 millimeters, the relative risk of dengue gradually increases but decreases when cumulative weekly rainfall exceeds 80 millimeters. Chien and Yu (2014) indicate a nonlinear association between rainfall and dengue fever. Generally, increased rainfall provides a conducive environment for mosquitoes, but extremely heavy rainfall can adversely affect them by washing away their habitats. With weekly maximum 24-hour rainfall of 50 millimeters or more, the risk of dengue fever increases. At the onset of rainfall, there is an increased risk of dengue fever, which can continue for at least three months. With extreme rainfall of 300 millimeters or more, dengue fever may be mitigated for one month.

Fouque and Reeder (2019) conclude that, with climate variability, the vectorial capacity will change, not the vectorial competence. Johns Hopkins Bloomberg School of Public Health defines vectorial capacity as "a measurement of the efficiency of vector-borne disease transmission," while "vector competence is an evaluation of the vector's capability (mechanical or biological) to transmit a pathogen and is, therefore, an additional component of vectorial capacity" (Norris 2006, 30). Fouque and Reeder (2019) conclude that (1) vectorial capacity is determined by vector density, which is dependent on rainfall patterns; (2) vector survival is related to temperature and humidity; (3) extrinsic incubation period of mosquitoes is related to temperature; and (4) biting behavior is determined genetically and also dependent on temperature. Their findings are corroborated by Shapiro et al. (2017) (figure A.1), who observe the rate of infectious bites increases with temperature as well as the vectorial capacity.

FIGURE A.1

Relationship between temperatures and infectious bites and vectorial capacity

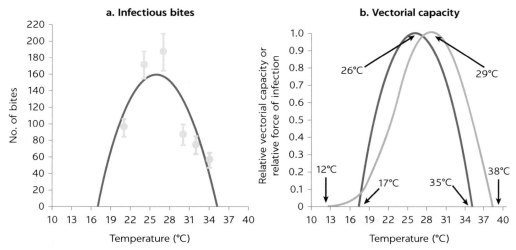

Source: Shapiro, Whitehead, and Thomas. 2017.
Note: Panel a: Best fit thermal performance curve for relative force of infection (here the number of infectious bites predicted for a cohort of 100 female mosquitoes); green points represent the calculated number of bites for the mean of both experimental blocks, and error bars represent standard deviations. *Panel b*: Comparison of scaled thermal performance curves for relative vectorial capacity and relative force of infection.

Ebi and Nealon (2016) identify globalization, trade, urbanization, travel, demographic change, inadequate domestic water supplies, and increasing temperatures as factors that affect the spread of main vectors of dengue. Based on data from Taiwan, China, Chien and Yu (2014) indicate that higher urbanized areas increase the risk of dengue transmission, which means locations with higher population density and socioeconomic status can be associated with increased incidence of dengue fever. According to Misslin et al. (2017), although urban areas are generally warmer than rural or subrural areas, there can be large differences in temperature within an urban area because of the creation of urban heat islands, which are areas that are significantly warmer than other areas because of human activities. Large urban areas can have a "mosaic of warm and cold areas," which can affect incidence of dengue fever cases—areas that are close to parks, for example, are cooler than built areas and the temperature differences can be as large as 7°C (Misslin et al. 2017). Ebi and Nealon (2016) indicate that further increases in temperature may reduce transmission of dengue virus in currently warm locations. They cite as an example data from Thailand over the period 1983 to 2010, where 80 percent of severe dengue cases occurred when the temperature was 27.0°C to 29.5°C and mean humidity was greater than 75 percent. They conclude that two most important factors for the incidence and distribution of dengue are daily mean temperature and the variation in temperature.

Fan and Liu (2019) project that an outbreak of dengue fever can shift from low-latitude areas to high latitude and from south to north in China. Their modeling is consistent with data from the Republic of Korea, Tanzania, and the United States. Fouque and Reeder (2019) mention that dengue disease is mostly urban and urban temperatures are drastically changing due to the warming climate, which is enhancing dengue transmission. They further conclude that lower mean temperatures at higher altitudes can increase dengue transmission, and they cite the example of increased dengue incidence in Nepal. They also indicate extreme weather events such as tropical cyclones are associated with increased dengue transmission, based on data from China. Chien and Yu (2014), however, indicate that extreme weather events that cause flooding can result in the elimination of mosquito breeding places and reduce the mosquito population. They also mention that other nonclimatic factors like immunity in humans and serotypes of dengue virus can be important compounders in the spread and dynamics of dengue fever.

REFERENCES

Ahmed, T. U., G. M. S. Rahman, K. Bashar, M. Shamsuzzaman, S. Samajpati, S. Sultana, M. I. Hossain, N. N. Banu, and M. S. Rahman. 2007. "Seasonal Prevalence of Dengue Vector Mosquitoes in Dhaka City, Bangladesh." *Bangladesh Journal of Zoology* 35 (2): 205–12.

Amin M. R., S. M. Tareq, and S. H. Rahman. 2011. "Impacts of Climate Change on Public Health: Bangladesh Perspective." *Global Journal of Environmental Research* 5 (3): 97–105.

Banu, S., W. Hu, Y. Guo, C. Hurst, and S. Tong. 2014. "Projecting the Impact of Climate Change on Dengue Transmission in Dhaka, Bangladesh." *Environment International* 63: 137–42.

Bashar, K., and N. Tuno. 2014. "Seasonal Abundance of *Anopheles* Mosquitoes and Their Association with Meteorological Factors and Malaria Incidence in Bangladesh." *Parasites & Vectors* 7: 442. https://parasitesandvectors.biomedcentral.com/articles/10.1186/1756-3305-7-442.

Berry, H., 2009. "Pearl in the Oyster: Climate Change as a Mental Health Opportunity." *Australian Psychiatry* 17 (6): 453–56.

Berry, H. L., K. Bowen, and T. Kjellstrom. 2010. "Climate Change and Mental Health: A Causal Pathways Framework." *International Journal of Public Health* 55 (2): 123–32.

Budiyono, R., S. P. Jati, and P. Ginandjar. 2017. "Potential Impact of Climate Variability on Respiratory Diseases in Infant and Children in Semarang." IOP Conference Series, *Earth Environmental Science* 55: 012049.

Caminade, C., S. Kovats, J. Rocklov, A. M. Tompkins, A. P. Morse, F. J. Colon-Gonzalez, H. Stenlund, P. Martens, and S. J. Lloyd. 2014. "Impact of Climate Change on Global Malaria Distribution." *Proceedings of the National Academy of Sciences* 111: 3286–91. doi:10.1073/pnas.1302089111.

Campbell-Lendrum, D., L. Manga, M. Bagayoko, and J. Sommerfeld. 2015. "Climate Change and Vector-Borne Diseases: What Are the Implications for Public Health Research and Policy?" *Philosophical Transactions of the Royal Society B* 370: 20130552. doi:10.1098/rstb.2013.0552.

Chien, L.-C., and H.-L. Yu. 2014. "Impact of Meteorological Factors on the Spatiotemporal Patterns of Dengue Fever Incidence." *Environmental International* 73: 46–56.

Choudhury, W. A., F. A. Quraishi, and Z. Haque. 2006. "Mental Health and Psychosocial Aspects of Disaster Preparedness in Bangladesh." *International Review of Psychiatry* 18 (6): 529–35. doi:10.1080/09540260601037896.

Chowdhury, F. R., Q. S. U. Ibrahim, M. S. Bari, M. M. J. Alam, S. J. Dunachie, A. J. Rodriguez-Morales, and M. I. Patwary. 2018. "The Association between Temperature, Rainfall, and Humidity with Common Climate-Sensitive Infectious Diseases in Bangladesh. *PLOS ONE* 13 (6): e0199579.

Colón-González, F. J., I. Lake, and P. R. Hunter. 2013. "The Effects of Weather and Climate Change on Dengue." PLOS *Neglected Tropical Diseases* 7 (11): e2503. doi:10.1371/journal.pntd.0002503.

Costa, E. A. P. de A., E. M. de M. Santos, J. C. Correia, and C. M. R. de Albuquerque. 2010. "Impact of Small Variations in Temperature and Humidity on the Reproductive Activity and Survival of *Aedes aegypti* (Diptera, Culcidae)." *Revista Brasileira de Entomologia* 54 (3): 488–93.

Costello, A., M. Abbas, A. Allen, S. Ball, S. Bell, R. Bellamy, S. Friel, et al. 2009. "Managing the Health Effects of Climate Change." *Lancet* 373: 1693–1733.

Ebi, K. L., and J. Nealon. 2016. "Dengue in a Changing Climate." *Environmental Research* 151 (2016): 115–23.

Fan, J. C., and Q. Y. Liu. 2019. "Potential Impacts of Climate Change on Dengue Fever Distribution Using RCP Scenarios in China." *Advances in Climate Change Research* 10: 1–8.

Flahault, A., R. R. de Castaneda, and I. Bolon. 2016. "Climate Change and Infectious Diseases." *Public Health Reviews* 37 (21): doi:10.1186/s40985-016-0035-2.

Fouque, F., and J. C. Reeder. 2019. "Impact of Past and On-going Changes on Climate and Weather on Vector-Borne Diseases Transmission: A Look at the Evidence." *Infectious Diseases of Poverty* 8 (51).

GOB (Government of Bangladesh). 2008. *Cyclone Sidr in Bangladesh: Damage, Loss and Needs Assessment for Disaster Recovery and Reconstruction*. Dhaka: GOB.

Hashizume, M., B. Armstrong, S. Hajat, Y. Wagatsuma, A. S. Faruque, T. Hayashi, and D. A. Sack. 2007. "Association between Climate Variability and Hospital Visits for Non-Cholera Diarrhea in Bangladesh: Effects and Vulnerable Groups." *International Journal of Epidemiology* 36 (5): 1030–37. doi:10.1093/ije/dym148.

Hashizume, M., A. M. Dewan, T. Sunahara, M. Z. Rahman, and T. Yamamoto. 2012. "Hydroclimatological Variability and Dengue Transmission in Dhaka, Bangladesh: A Time-Series Study." *BMC Infectious Diseases* 12 (98). http://www.biomedcentral.com/1471-2334/12/98.

Hay, S. I., A. J. Tatem, C. A. Guerra, and R. W. Snow. 2006. *Foresight on Population at Malaria Risk in Africa: 2005, 2015 and 2030*. London: Foresight Project, Office of Science and Innovation.

Hossain, M., E. Noiri, and K. Moji. 2011. "Climate Change and Kala-azar." In *Kala-azar in South Asia*, edited by T. Jha and E. Noiri, 127–37. Dordrecht, Netherlands: Springer. doi:10.1007/978-94-007-0277-6_12.

Islam, M. Z., S. Rutherford, D. Phung, M. N. Uzzzaman, S. Baum, M. M. Huda, M. Asaduzzaman, M. R. R. Talukder, and C. Chu. 2018. "Correlates of Climate Variability and Dengue Fever in Two Metropolitan Cities in Bangladesh." *Cureus*. 10 (10): e3398. doi:10.7759/cureus.3398.

Kabir, S. M. S. 2018. "Psychological Health Challenges of the Hill-Tracts Region for Climate Change in Bangladesh." *Asian Journal of Psychiatry* 34 (2018) 74–77.

Kabir, M. I., M. B. Rahman, W. Smith, M. A. F. Lusha, and A. H. Milton. 2016. "Climate Change and Health in Bangladesh: A Baseline Cross-Sectional Survey." *Global Health Action* 9 (1): 29609. doi:10.3402/gha.v9.29609.

Kakarla, S. G., C. Caminade, S. R. Mutheneni, A. P. Morse, S. M. Upadhyayula, M. R. Kadiri, and S. Kumaraswamy. 2019. "Lag Effect of Climatic Variables on Dengue Burden in India." *Epidemiology & Infection* 147: e170. doi:10.1017/S0950268819000608. https://pubmed.ncbi .nlm.nih.gov/31063099/.

Khan, W. A., S. R. Galagan, C. S. Prue, J. Khyang, S. Ahmed, M. Ram, M. S. Alam, et al. 2014. "Asymptomatic *Plasmodium falciparum* Malaria in Pregnant Women in the Chittagong Hill Districts of Bangladesh." *PLOS One* 9 (5): e98442. doi.10.1371/journal.pone.0098442.

Kunii, O., S. Nakamur, R. Abdur, and S. Waka. 2002. "The Impact on Health and Risk Factors of the Diarrhoea Epidemics in the 1998 Bangladesh Floods." *Public Health* 116: 68–74.

Lau, C. L., L. D. Smythe, S. B. Craig, and P. Weinstein. 2010. "Climate Change, Flooding, Urbanisation, and Leptospirosis: Fuelling the Fire?" *Transactions of the Royal Society of Tropical Medicine and Hygiene* 104: 631–38.

Lowe, R., A. M. Stewart-Ibarra, D. Petrova, M. Garcia-Díez, M. J. Borbor-Cordova, R. Mejía, M. Regato, and X. Radó. 2017. "Climate Services for Health: Predicting the Evolution of the 2016 Dengue Season in Machala, Ecuador." *Lancet Planet Health* 1: e142– e151.

Mahmud, A. S., M. I. Kabir, K. Engo-Monsen S. Tahmina, B. K. Riaz, M. A. Hossain, F. Khanom, et al. 2019. "Megacities as Drivers of National Outbreaks: The Role of Holiday Travel in the Spread of Infectious Diseases." *bioRxiv*, August 16, 2019. doi:10.1101/737379.

McMichael, T. 2012. "Health Risks, Present and Future, from Global Climate Change." *BMJ* 344: e1359. doi:10.1136/bmj.e1359.

McMichael, A. J., R. E. Woodruff, and S. Hales. 2006. "Climate Change and Human Health: Present and Future Risks." *Lancet* 367 (9513): 859–69.

Mirsaeidi, M., H. Motahari, M. T. Khamesi, A. Sharifi, M. Campos, and D. E. Schraufnagel. 2015. "Climate Change and Respiratory Infections." *Annals of the American Thoracic Society* 13 (8).

Misslin, R., O. Telle, E. Daudé, A. Vaguet, and R. E. Paul. 2017. "Urban Climate versus Global Climate Change: What Makes the Difference for Dengue?" *Annals of the New York Academy of Sciences* 1382 (1): 56–72.

Mone, F. H. 2019. "Sustainable Actions Needed to Mitigate Dengue Outbreak in Bangladesh." *Lancet* 19.

Mutsuddy, P., S. T. Jhora, A. K. M. Shamsuzzaman, S. M. G. Kaisar, and M. N. A. Khan. 2018. "Dengue Situation in Bangladesh: An Epidemiological Shift in Terms of Morbidity and Mortality." *Canadian Journal of Infectious Diseases and Medical Microbiology* 2019, article ID 3516284. doi:10.1155/2019/3516284.

Nabi, S. A., and S. S. Qader. 2009. Is Global Warming Likely to Cause an Increased Incidence of Malaria? *Libyan Journal of Medicine* 2009 (4): 9–16.

Nahar, N., Y. Blomstedt, B. Wu, I. Kandarina, L. Trisnantoro, and J. Kinsman. 2014. "Increasing the Provision of Mental Health Care for Vulnerable, Disaster-Affected People in Bangladesh." *BMC Public Health* 14 (708). doi:10.1186/1471-2458-14-708.

Norris, Douglas E. 2006. "Malaria Entomology." Presentation. Johns Hopkins Bloomberg School of Public Health. https://www.glowm.com/pdf/M-lecture4.pdf.

Obradovich, N., R. Migliorini, M. P. Paulus, and I. Rahwan. 2018. "Empirical Evidence of Mental Health Risk Posed by Climate Change." *Proceedings of the National Academy of Sciences* 115 (43): 10953–58.

Ogden, N. H. 2017. "Climate Change and Vector-Borne Diseases of Public Health Significance." *FEMS Microbiology Letters* 364 (19). doi:10.1093/femsle/fnx186.

Ooi, E. E., A. Wilder-Smith, L. C. Ng, and D. J. Gubler. 2010. "The 2007 Dengue Outbreak in Singapore." *Epidemiology and Infec*tion 138: 958–59, author reply 9–61. doi:10.1017 /S0950268810000026.

Page, L. A., and L. M. Howard. 2010. "The Impact of Climate Change on Mental Health (But Will Mental Health Be Discussed at Copenhagen?)" *Psychological Medicine* 40 (2): 177–80.

Patwary, M. I., M. Z. I. Bari, and T. Islam. 2016. "Epidemiological and Clinical Importance of Leptospirosis: Bangladesh Perspective." *Bangladesh Medical Journal* 45 (3).

Patz, G. A., J. P. McCarty, S. Hussein, U. Confalonieri, and N. D. Wet. 2003. "Climate Change and Infectious Diseases." In *Climate Change and Human Health: Risks and Responses,* edited by A. J. McMichael, D. H. Campbell-Lendrum, C. F. Corvalán, K. L. Ebi, A. K. Githeko, J. D. Scheraga, and A. Woodward, 103–32. Geneva: World Health Organization.

Paul, B. K. 2009. "Why Relatively Fewer People Died? The Case of Bangladesh's Cyclone Sidr." *Natural Hazards* 50: 289–304. doi:10.1007/s11069-008-9340-5.

Phakhounthong, K., P. Chaovalit, P. Jittamala, S. D. Blacksell, M. J. Carter, P. Turner, K. Chheng et al. 2018. "Predicting the Severity of Dengue Fever in Children on Admission Based on Clinical Features and Laboratory Indicators: Application of Classification Tree Analysis." *BMC Pediatrics* 18 (1): 109.

Raheel, U., M. Faheem, M. N. Riaz, N. Kanwal, F. Javed, N. S. S. Zaidi, and I. Qadri. 2010. "Dengue Fever in the Indian Subcontinent: An Overview." *Journal of Infection in Developing Countries* 5 (4): 239–47.

Rahman, M. M., S. Ahmad, A. S. Mahmud, M. Hassan-uz-Zaman, M. A. Nahian, A. Ahmed, Q. Nahar, and P. K. Streatfield. 2019. "Health Consequences of Climate Change in Bangladesh: An Overview of the Evidence, Knowledge Gaps and Challenges." *WIREs Climate Change* 10 (5): e601. doi:10.1002/wcc.601.

Rana, S., and S. Singh. 2015. "Nipah Virus: Effects of Urbanization and Climate Change." Paper prepared for Third International Conference on Biological, Chemical, and Environmental Sciences, Kuala Lumpur, September 21–22, 2015.

Reid, H. L., U. Haque, S. Roy, N. Islam, and A. C. Clements. 2012. "Characterizing the Spatial and Temporal Variation of Malaria Incidence in Bangladesh, 2007." *Malaria Journal* 11 (170): 170. doi:10.1186/1475-2875-11-170.

Rylander, C., J. O. Odland, and T. M. Sandanger. 2013. "Climate Change and the Potential Effects on Maternal and Pregnancy Outcomes: An Assessment of the Most Vulnerable—the Mother, Fetus, and Newborn" *Global Health Action* 6: 19538. doi:10.3402/gha.v6i0.19538.

Sadoine, M. L., A. Smargiassi, V. Ridde, L. C. Tusting, and K. Zinszer. 2018. "The Associations between Malaria, Interventions, and the Environment: A Systemic Review and Meta-Analysis." *Malaria Journal* 17 (73). doi:10.1186/s12936-018-2220-x.

Schwartz, B. S., J. B. Harris, A. I. Khan, R. C. Larocque, D. A. Sack, M. A. Malek, A. S. G. Faruque, et al. 2006. "Diarrheal Epidemics in Dhaka, Bangladesh, during Three Consecutive Floods: 1988, 1998, and 2004." *American Society of Tropical Medicine and Hygiene* 74 (6): 1067–73.

Shapiro, L .L. M., S. A. Whitehead, and M. B. Thomas. 2017. "Quantifying the Effects of Temperature on Mosquito and Parasite Traits That Determine the Transmission Potential of Human Malaria." *PLOS Biology* 15 (10): e2003489. doi:10.1371/journal. pbio.2003489.

Sharmin, S., K. Glass, E. Viennet, and D. Harley. 2015. "Interaction of Mean Temperature and Daily Fluctuation Influences Dengue Incidence in Dhaka, Bangladesh." *PLOS Neglected Tropical Diseases* 9 (7): e0003901. doi:10.1371/journal.pntd.0003901.

Siddique, A. K., Q. Islam, K. Akram, Y. Mazumder, A. Mitra, and A. Eusof. 1989. "Cholera Epidemic and Natural Disasters; Where Is the Link." *Tropical and Geographical Medicine* 41: 377–82.

Sirisena, P. D. N. N., and F. Noordeen. 2013. "Evolution of Dengue in Sri Lanka—Changes in the Virus, Vector, and Climate." *International Journal of Infectious Diseases* 19: 6–12.

Swim, J., S. Clayton, T. Doherty, R. Gifford, G. Howard, J. Reser, P. Stern, and E. Weber. 2009. *Psychology and Global Climate Change: Addressing a Multifaceted Phenomenon and Set of Challenges.* Washington, DC: American Psychological Association. www.apa.org/science /about/publications/climate-change.aspx.

Takaro, T. K., K. Knowlton, and J. R. Balmes. 2013. "Climate Change and Respiratory Health: Current Evidence and Knowledge Gaps" *Expert Review of Respiratory Medicine* 7 (4): 349–61.

Tauil, P. L. 2001. "Urbanização e ecologia do dengue" (Urbanization and dengue ecology). *Cad Saude Publica* 2001 (17): suppl. 99–102. Portuguese. https://pubmed.ncbi.nlm.nih .gov/11426270.

WEDO (Women's Environment and Development Organization). 2008. *Gender, Climate Change and Human Security: Lessons from Bangladesh, Ghana, and Senegal.* New York, WEDO.

WHO (World Health Organization). 2002. *Gender and Health in Disasters*, Geneva: WHO.

WHO (World Health Organization). 2019. "Dengue and Severe Dengue." Geneva: WHO. https://www.who.int/news-room/fact-sheets/detail/dengue-and-severe-dengue.

Wu, X., Y. Lu, S. Zhou, L. Chen, and B. Xu. 2016. "Impact of Climate Change on Human Infectious Diseases: Empirical Evidence and Human Adaptation." *Environment International* 86, 14–23. doi:10.1016/j.envint.2015.09.007.

Wu, J., M. Yunus, M. Ali, and M. E. Escamilla. 2018. "Influences of Heatwave, Rainfall, and Tree Cover on Cholera in Bangladesh." *Environment International* 120 (November): 304–11.

Suplementary Tables on Demography, Socioeconomic Characteristics, and Disease Patterns, by Location, in Bangladesh

TABLE B.1 **Demographic characteristics of the sample at baseline**

	NATIONAL		URBAN ALL		URBAN DHAKA AND CHATTOGRAM		RURAL		TEST	
	MEAN	SD	MEAN	SD	MEAN	SD	MEAN	SD	URBAN–RURAL	CITIES–RURAL
VARIABLES	(1)	(2)	(3)	(4)	(5)	(6)	(7)	(8)	(9)	(10)
Male/female (1/0)	0.50	0.50	0.50	0.50	0.52	0.50	0.50	0.50	n.a.	n.a.
Age	28.00	19.50	28.30	18.90	27.80	17.50	27.90	19.60	n.a.	n.a.
Marital status										
Married (1/0)	0.51	0.50	0.51	0.50	0.49	0.50	0.51	0.50	n.a.	n.a.
Never married (1/0)	0.23	0.42	0.24	0.43	0.26	0.44	0.23	0.42	n.a.	*
Other (1/0)	0.26	0.44	0.25	0.43	0.24	0.43	0.26	0.44	n.a.	n.a.
Years of education	4.92	4.46	5.98	4.91	6.06	5.08	4.64	4.30	***	***
Sex of hh head male/female (1/0)	0.92	0.28	0.91	0.28	0.88	0.33	0.92	0.28	n.a.	***
Age of hh head	46.00	13.30	46.30	13.10	44.30	12.80	46.20	13.40	n.a.	***
Years of education of hh head	4.51	4.69	5.80	5.32	6.07	5.66	4.18	4.45	***	***
Individual member with a disability (1/0)	0.13	0.34	0.14	0.35	0.15	0.36	0.13	0.34	n.a.	n.a.

N = 15,383

Source: Original table for this publication.

Note: Table shows weighted means. Tests (columns 9 and 10) show significance levels from a weighted *t*-test: *p < 0.10; **p < 0.05; ***p < 0.01. hh = household; n.a. = not applicable; SD = standard deviation.

TABLE B.2 **Socioeconomic characteristics of the sample at baseline**

VARIABLES	NATIONAL		URBAN				RURAL		TEST	
			ALL		DHAKA AND CHATTOGRAM CITIES					
	MEAN	SD	MEAN	SD	MEAN	SD	MEAN	SD	URBAN–RURAL	CITIES–RURAL
	(1)	(2)	(3)	(4)	(5)	(6)	(7)	(8)	(9)	(10)
Wall material										
Straw/mud	0.19	0.34	0.04	0.14	0.00	0.00	0.22	0.37	***	***
Tin	0.47	0.50	0.32	0.46	0.13	0.33	0.51	0.50	***	***
Brick/cement	0.34	0.48	0.64	0.48	0.87	0.33	0.27	0.44	***	***
Roof material										
Tin	0.85	0.35	0.68	0.47	0.48	0.50	0.90	0.30	***	***
Brick/cement	0.13	0.34	0.31	0.46	0.52	0.50	0.08	0.28	***	***
Has access to electricity (1/0)	0.88	0.32	0.98	0.13	1.00	0.06	0.86	0.35	***	***
Stove type: clean (1) / unclean (0)	0.20	0.40	0.56	0.50	0.97	0.17	0.10	0.30	***	***
Rooms per capita	0.54	0.28	0.54	0.31	0.46	0.31	0.54	0.27	n.a.	***
Has separate dining room (1/0)	0.29	0.45	0.27	0.44	0.21	0.41	0.30	0.46	***	***
Asset quintiles (Q)										
Q1: Poorest	0.24	0.43	0.10	0.30	0.06	0.24	0.28	0.45	***	***
Q2	0.23	0.42	0.14	0.35	0.08	0.27	0.25	0.43	***	***
Q3	0.20	0.40	0.17	0.38	0.15	0.35	0.21	0.41	***	***
Q4	0.17	0.38	0.23	0.42	0.26	0.44	0.15	0.36	***	***
Q5: Richest	0.16	0.37	0.35	0.48	0.45	0.50	0.11	0.32	***	***

N = 15,383

Source: Original table for this publication.
Note: Table shows weighted means. Tests (columns 9 and 10) show significance levels from a weighted *t*-test: *$p < 0.10$; **$p < 0.05$; ***$p < 0.01$. n.a. = not applicable; SD = standard deviation.

TABLE B.3 **Infectious diseases across locations**

VARIABLES	NATIONAL		ALL		DHAKA AND CHATTOGRAM CITIES		RURAL		TEST	
			URBAN							
	MEAN	SD	MEAN	SD	MEAN	SD	MEAN	SD	URBAN-RURAL	CITIES-RURAL
	(1)	(2)	(3)	(4)	(5)	(6)	(7)	(8)	(9)	(10)
Any infectious disease (1/0)	0.06	0.24	0.06	0.24	0.08	0.27	0.06	0.23	n.a.	**
Waterborne disease (1/0)	0.14	0.35	0.13	0.34	0.19	0.40	0.14	0.35	n.a.	n.a.
Respiratory disease (1/0)	0.64	0.48	0.65	0.48	0.50	0.50	0.63	0.48	n.a.	*
Vector-borne disease (1/0)	0.22	0.42	0.22	0.42	0.30	0.46	0.22	0.42	n.a.	n.a.
Individual had a common cold (1/0)	0.09	0.29	0.09	0.29	0.06	0.25	0.10	0.29	n.a.	***

N = 15,383

Source: Original table for this publication.

Note: Table shows weighted means. Waterborne, respiratory, and vector-borne diseases are conditional on whether an individual reported an infectious disease (excluding common cold) in the four weeks preceding the survey. Whether an individual had a common cold represents the sample population. Tests (columns 9 and 10) show significance levels from a weighted t-test: $^* p < 0.10$; $^{**} p < 0.05$; $^{***} p < 0.01$. n.a. = not applicable; SD = standard deviation.